Kroc Foundation Series
Volume 16

WHITE CELL MECHANICS:
BASIC SCIENCE AND CLINICAL ASPECTS

Proceedings of a symposium held at the Kroc Foundation,
Santa Ynez Valley, California, May 2–6, 1983

Editors

Herbert J. Meiselman
Department of Physiology and Biophysics
University of Southern California School of Medicine
Los Angeles, California

Marshall A. Lichtman
Hematology Unit
Department of Medicine
University of Rochester School of Medicine and Dentistry
Rochester, New York

Paul L. LaCelle
Department of Radiation Biology and Biophysics
University of Rochester School of Medicine and Dentistry
Rochester, New York

Alan R. Liss, Inc. • **New York**

Address all Inquiries to the Publisher
Alan R. Liss, Inc., 150 Fifth Avenue, New York, NY 10011

Library of Congress Cataloging in Publication Data

Main entry under title:

White cell mechanics.

(Kroc Foundation series; v. 16)
Includes bibliographical references and index.
1. Leucocytes--Congresses. 2. Biomechanics--Congresses. 3. Cell interaction--Congresses. 4. Leucocyte disorders--Congresses.   I. Meiselman, Herbert J.
II. Lichtman, Marshall A.   III. LaCelle, Paul L.
IV. Series. [DNLM: 1. Leukocytes--Congresses. W3 KR45 v.16 / WH 200 W589 1983]
QP95.W45   1983   616.1′54   83-19968
ISBN 0-8451-0306-7

Contents

I. CELLULAR AND MEMBRANE MECHANICS OF WHITE CELLS

II. WHITE CELL CYTOPLASMIC AND MEMBRANE PROPERTIES

III. WHITE CELL FLOW BEHAVIOR

IV. WHITE CELL LOCOMOTION AND INTERACTION WITH ENDOTHELIUM

V. THE CONTRIBUTION OF WHITE CELLS TO PATHOLOGIC FLOW STATES

Contributors

J.A. Badwey, Department of Biological Chemistry, Harvard University Medical School, Boston, MA 02115 **[111]**

Ulf Bagge, Laboratory of Experimental Biology, Department of Anatomy, University of Göteborg, Göteborg, Sweden **[285]**

C.B. Berde, Department of Pediatrics, Harvard University Medical School, Boston, MA 02115 **[111]**

Anthony T.W. Cheung, Department of Pediatrics, School of Medicine, and California Primate Research Center, University of California at Davis, Davis, CA 95616 **[169]**

Shu Chien, Department of Physiology, Columbia University College of Physicians & Surgeons, New York, NY 10032 **[3,19]**

Yee-Hon Chin, Department of Pathology, SUNY Downstate Medical Center, Brooklyn, NY 11203 **[255]**

Alfred Clark, Jr., Department of Mechanical Engineering, University of Rochester, Rochester, NY 14642 **[221]**

Patricia Clark, Department of Mathematics, Rochester Institute of Technology, Rochester, NY 14623 **[221]**

G.F. Clough, Department of Pharmacology, Institute of Basic Medical Sciences, Royal College of Surgeons of England, Lincoln's Inn Fields, London WC2A 3PN, England **[195]**

J.T. Curnutte, Department of Pediatrics, Harvard University Medical School, Boston, MA 02115 **[111]**

Mark D. Dahlgren, Department of Medicine, Veterans Administration Medical Center, University of California at San Diego, La Jolla, CA 92093 **[271]**

Robert L. Engler, Department of Medicine, Veterans Administration Medical Center, University of California at San Diego, La Jolla, CA 92093 **[271]**

Suzanne G. Eskin, Department of Surgery, Baylor College of Medicine, Houston, TX 77030 **[209]**

Evan A. Evans, Department of Pathology, University of British Columbia, Vancouver, British Columbia, Canada V6T 1W5 **[53]**

M.J. Forrest, Department of Pharmacology, Institute of Basic Medical Sciences, Royal College of Surgeons of England, Lincoln's Inn Fields, London WC2A 3PN, England **[195]**

P. Gaehtgens, Institut für Physiologie der Freien Universitat Berlin, 1000 Berlin 33, West Germany **[147,159]**

John I. Gallin, LCI/NIAID, National Institutes of Health, Bethesda, MD 20205 **[95]**

Harry L. Goldsmith, McGill University Medical Clinic, Montreal General Hospital, Montreal, Quebec, Canada H3G 1A4 **[131]**

The bold face numbers are opening page numbers of each author's article.

Lewis L. Hsu, Department of Radiation Biology and Biophysics, University of Rochester School of Medicine and Dentistry, Rochester, NY 14642 **[221]**

P.J. Jose, Department of Pharmacology, Institute of Basic Medical Sciences, Royal College of Surgeons of England, Lincoln's Inn Fields, London WC2A 3PN, England **[195]**

M.J. Karnovsky, Department of Pathology, Harvard University Medical School, Boston, MA 02115 **[111]**

Manfred L. Karnovsky, Department of Biological Chemistry, Harvard University Medical School, Boston, MA 02115 **[111]**

Donald L. Kreutzer, Department of Pathology, University of Connecticut Health Center, Farmington, CT 06032 **[87]**

Paul L. LaCelle, Department of Radiation Biology and Biophysics, University of Rochester School of Medicine and Dentistry, Rochester, NY 14642 **[xi]**

J. M. Lackie, Department of Cell Biology, Glasgow University, Glasgow, G12 8QQ, Scotland **[237]**

Marshall A. Lichtman, Hematology Unit, Department of Medicine, University of Rochester School of Medicine and Dentistry, Rochester, NY 14642 **[xi,295]**

Larry V. McIntire, Biomedical Engineering Laboratory, Rice University, Houston, TX 77251 **[209]**

Herbert J. Meiselman, Department of Physiology and Biophysics, University of Southern California School of Medicine, Los Angeles, CA 90033 **[xi]**

Michael E. Miller, Department of Pediatrics, School of Medicine, and California Primate Research Center, University of California at Davis, Davis, CA 95616 **[169]**

Jayasree Nath, LCI/NIAID, National Institutes of Health, Bethesda, MD 20205 **[95]**

U. Nobis, Institut für Normale und Pathologische Physiologie der Universität, 5000 Köln-Lindenthal, West Germany **[147]**

Michael A. Peterson, Department of Medicine, Veterans Administration Medical Center, University of California at San Diego, La Jolla, CA 92093 **[271]**

Thomas D. Pollard, Department of Cell Biology and Anatomy, Johns Hopkins Medical School, Baltimore, MD 21205 **[75]**

A.R. Pries, Institut für Physiologie der Freien Universität Berlin, 1000 Berlin 33, West Germany **[147]**

J.M. Robinson, Department of Pathology, Harvard University Medical School, Boston, MA 02115 **[111]**

E.A. Schmalzer, Department of Physiology, Columbia University College of Physicians & Surgeons, New York, NY 10032 **[19]**

Geert W. Schmid-Schönbein, Department of Applied Mechanics and Engineering Sciences, University of California at San Diego, La Jolla, CA 92093 **[3,19,271]**

Michael P. Sheetz, Department of Physiology, University of Connecticut Health Center, Farmington, CT 06032 **[87]**

Richard Skalak, Department of Civil Engineering and Engineering Mechanics, Bioengineering Institute, Columbia University, New York, NY 10027 **[3,19]**

Samira Spain, McGill University Medical Clinic, Montreal General Hospital, Montreal, Quebec, Canada H3G 1A4 **[131]**

K.-L.P. Sung, Department of Physiology, Columbia University College of Physicians & Surgeons, New York, NY 10032 **[19]**

Wen-Pin Wang, Department of Pathology, University of Connecticut Health Center, Farmington, CT 06032 [87]

Richard E. Waugh, Department of Radiation Biology and Biophysics, University of Rochester School of Medicine and Dentistry, Rochester, NY 14642 [221]

C.V. Wedmore, Department of Pharmacology, Institute of Basic Medical Sciences, Royal College of Surgeons of England, Lincoln's Inn Fields, London WC2A 3PN, England [195]

P.C. Wilkinson, Department of Bacteriology and Immunology, Glasgow University, Glasgow, G11 6NT, Scotland [237]

T.J. Williams, Department of Pharmacology, Institute of Basic Medical Sciences, Royal College of Surgeons of England, Lincoln's Inn Fields, London WC2A 3PN, England [195]

Judith J. Woodruff, Department of Pathology, SUNY Downstate Medical Center, Brooklyn, NY 11203 [255]

WHITE CELL MECHANICS: BASIC SCIENCE AND CLINICAL ASPECTS

The Kroc Foundation   May 2-6, 1983

First Row: Donald Whedon, Robert Kroc, Marshall Lichtman, Herbert Meiselman, Judith Woodruff, Harold Wayland, Anthony Cheung, Shu Chien, Richard Skalak
Second Row: Geert Schmid-Schönbein, Evan Evans, Richard Waugh, Ulf Bagge, Robert Hochmuth, Peter Gaehgens, Donald McMillan, Harry Goldsmith, John Gallin, Thomas Pollard
Third Row: Gerard Nash, Manfred Karnovsky, Walter Garey, Larry McIntire, John Lackie, Timothy Williams, Michael Sheetz

Preface

For many years the focus of blood rheologists and, in fact, circulatory physiologists has been on red blood cells, their unique mechanical properties, and their key role in gas transfer. Initially, the contribution of red cells to the overall rheologic properties of blood and the influence of shear rate and plasma protein composition on the viscosity of blood were studied extensively. Subsequently, the mechanical behavior of the red cell itself became the subject of extensive analyses. The physical properties of normal and abnormal red cells (eg, size, shape, surface area, hemoglobin content) as well as mechanical properties such as cellular deformability, cellular viscoelasticity, and membrane mechanical behavior have been analyzed in extenso and treated quantitatively. Newer techniques were added to those of viscometry, including micropore filterability, micropipette aspiration, and the observation of red cell shape in various well-defined fluid shear fields.

Recently rheologists, biophysicists, and clinicians have begun to examine the mechanical properties of white cells, their behavior in the circulation, and their role in the pathogenesis of certain abnormal phenomena. It has been known for decades that normal neutrophils produce temporary obstructions in capillaries, which lead to a transient cessation of blood flow in those vessels. Since leukocytes normally occupy less than one percent of the volume of blood as compared to the 45% occupied by red cells in the healthy individual, there has been little prior concern with the possible influence of the bulk of white cells on blood flow and oxygen transport. Within the last few years, however, considerable interest has been generated in the effects of individual white cells on microcirculatory flow dynamics. The results of these studies indicate that white cells are far more resistant to deformation than red cells, that they tend to attach to surfaces much more readily than normal RBC, and that generalizations referable to leukocyte behavior cannot be made. Different leukocyte types have different physical and chemical properties and circulatory patterns. Thus, the circulation of lymphocytes and neutrophils is neither similar nor likely to be totally related to common chemical or physical properties of the cell or its membrane.

Because of the complexities inherent in studies of the mechanical properties of white cells, divergent interest groups were asked to share their techniques and knowledge relevant to these cells. The forum for this interchange was a conference on "White Cell Mechanics: Basic Science and Clinical Aspects" held at the Kroc Foundation in the Santa Ynez Valley, California, from May 2–6, 1983. When the editors convened to discuss the format of this conference, we decided that we would bring together participants with interests in selected areas to provide cross-fertilization within the time constraints of the meeting. We elected to consider the mechanical properties of the white cell during passive and active movement, the influence of their mechanical properties on the flow characteristics within the circulation or in artificial channels, the behavior of leukocytes during active motion in systems simulating extravascular or tissue spaces, and the influence of leukocytes in the pathogenesis of abnormal flow conditions. To accomplish this we needed to integrate theoreticians and experimentalists, including biophysicists, engineers, microvascular physiologists, cell physiologists, and cellular biochemists so as to synthesize hypotheses of passive and active leukocyte behavior when viewed from studies on whole cells or subcellular constituents and as examined by different techniques.

The formal presentations at the conference generated considerable informal discussion, much of it unusually informative because it crossed disciplinary and research area boundaries. The interdisciplinary interactions generated in this diverse group of participants were both thought provoking and beneficial. Given the breadth of interests of the group, we believe that this resulting monograph represents a comprehensive description of the state of knowledge of white cell mechanics. It is our hope that this publication will foster and facilitate research on the mechanical properties of white cells and, in doing so, will enhance our knowledge of their contributions to patterns of flow in the microcirculation and to their behavior in the tissues.

The hospitality of Dr. and Mrs. Robert Kroc and the staff of the Kroc Foundation was extraordinary! We join with the other participants in expressing our sincere gratitude to the Kroc Foundation for having made this a highly successful meeting and for giving us the opportunity to disseminate the information more widely through the publication of this monograph as a volume of the Kroc Foundation Series. Lastly, the editors wish to thank all of the participants for taking time from their busy schedules to attend this meeting–without them this conference would still be a "paper dream" of the editors.

Herbert J. Meiselman
Marshall A. Lichtman
Paul L. LaCelle

I. CELLULAR AND MEMBRANE MECHANICS OF WHITE CELLS

White Cell Mechanics: Basic Science and
Clinical Aspects, pages 3–18

Viscoelastic Deformation of White Cells: Theory and Analysis

Richard Skalak, Shu Chien, and Geert W. Schmid-Schönbein

*Department of Civil Engineering, Bioengineering Institute, (R.S.) and
Department of Physiology, College of Physicians & Surgeons (S.C.),
Columbia University, New York, New York 10027 and
AMES-Bioengineering, University of California, San Diego, La Jolla,
California 92093 (G.W.S.-S.)*

INTRODUCTION

The rheological behavior of leukocytes has been known for some time to be viscoelastic, comparatively stiff, and highly viscous compared to red blood cells. This comparison can be drawn from observations made directly in the in vivo microcirculation in man [Brånemark, 1971]. More quantitative experiments in which leukocytes are drawn into tapering glass capillaries [Bagge et al, 1977a,b] or micropipettes [Schmid-Schönbein et al, 1981; LaCelle et al, 1982] confirm the impression of viscoelastic behavior. In addition to the passive viscoelastic behavior of leukocytes in response to applied deformations or stresses, white blood cells exhibit active motion in various situations of free suspension [Schmid-Schönbein et al, 1982], migration through the capillary wall and the interstitium, locomotion along a surface and in deformation during phagocytosis [Dembo et al, 1981]. During active motions, leukocytes project pseudopods which have been shown to be stiffer than the passive white cell cytoplasm. It has been demonstrated that such pseudopods contain actin networks [Stossel, 1982] and that processes of gelation and solation are involved.

In the subsequent sections, a summary of the current theories of the rheology of the active and passive behavior of leukocytes will be given [Skalak et al, 1982; Schmid-Schönbein et al, 1983]. The passive properties of leukocytes have been extensively tested by the micropipette technique and will be reported for various conditions of osmolarity, temperature, and pH in the following companion paper. The theory of active motion is still under development, and a particular proposal based on the gelation front being the source of the active motion will be presented.

THEORY OF PASSIVE LEUKOCYTE DEFORMATION

Experiments show that when a stress is suddenly applied to a leukocyte, it responds with an initial deformation immediately and then continues to deform more slowly. These characteristics can be supplied by a so-called Maxwell model which contains an elastic spring in series with a viscous element. Experiments also show that after release of stress or deformation the white blood cell in free suspension returns to its initial spherical shape. A simple model which can reproduce both the loading and recovery behavior is shown in Figure 1. The spring k_1 provides the restoring force necessary to recover the initial shape after release of stress.

The model shown in Figure 1 is intended to represent the leukocyte behavior as a whole under small deformations. At this time, the identification of the elements of the model with specific components of the cell is not precisely or completely established. However, it is clear that the principal viscoelastic elements (k_2 and μ) are associated with the cytoplasm of the cell. This is in contrast to the red blood cell, in which it is well known that the membrane and its associated cytoskeleton provide the principal elasticity of the cell. In the model of the leukocyte developed here, the membrane is at first neglected. The cell is modeled as a uniform viscoelastic sphere having the constitutive characteristics shown in Figure 1. The membrane of leukocytes is convoluted when it is in a roughly spherical state, and, as will be shown in the next section, very small forces are required to produce unfolding of the membrane itself. Once the folds of the membrane are stretched out, the membrane becomes very stiff against any changes in surface area. This has been demonstrated by swelling white blood cells to smooth spheres in hypotonic solution and evaluating the isotropic modulus of the membrane. In this state, it is similar to the red blood cell membrane in resisting strongly any change of area.

The restoring elastic element k_1 in Figure 1 may be due to the action of the membrane plus some cytoskeletal layer near the surface of the cell. Such a layer containing a small amount of myosin [Stossel and Pollard, 1973] could by random active contraction draw the cell back into a spherical shape slowly after a large deformation. It is also possible that the Brownian motion of the granules and cytoplasm of the cell play some role in the gradual restoration of the spherical shape of the cell.

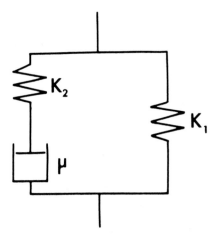

Fig. 1. Viscoelastic solid model of leukocyte behavior.

In the present model the presence of the granules and nucleus of the cell is not taken specifically into account although they may be contributing to the viscoelastic properties k_2, μ of the model. A more accurate model should consist of a viscoelastic nucleus enclosed in its own membrane and having properties which are stiffer than the cytoplasm of the cell. The rheological influence of the membrane of the nucleus can probably be disregarded as long as it is not stretched taut. The presence of granules may also be neglected as specific entities in passive deformations. Their influence will contribute to the constants k_2 and μ of the model.

The model shown in Figure 1 is interpreted as a three-dimensional continuum constitutive equation representing the shear behavior of the cell. The volumetric behavior of each of the constituents of the cell is assumed to be incompressible. Then the conservation of mass requires

$$e_{ij} = 0, \tag{1}$$

where e_{ij} is the strain tensor defined by

$$e_{ij} = \frac{1}{2}\left(\frac{\partial u_j}{\partial x_i} + \frac{\partial u_i}{\partial x_j}\right), \tag{2}$$

in which u_i is the displacement from the initial unstressed position. In the index notation used [Fung, 1965], i and j may take the values 1, 2, 3 and the repeated index as in Equation 1 indicates summation, ie, $e_{ii} = e_{11} + e_{22} + e_{33}$. For incompressible materials, the mean stress is represented by a pressure p which does not depend on the strain and is defined by

$$p = -\frac{1}{3} \sigma_{ii}, \tag{3}$$

where σ_{ij} is the stress tensor. The shear behavior involves the stress and strain deviators σ'_{ij} and e'_{ij} defined by

$$\sigma'_{ij} = \sigma_{ij} - \frac{1}{3} \sigma_{kk}\delta_{ij} \qquad e'_{ij} = e_{ij} - \frac{1}{3} e_{kk}\delta_{ij}. \tag{4}$$

In the incompressible case assumed here, e_{kk} in Equation 4 is zero by Equation 1, so $e'_{ij} = e_{ij}$. The model shown in Figure 1 leads to the three-dimensional constitutive equation

$$\sigma'_{ij} + \frac{\mu}{k_2} \frac{\partial \sigma'_{ij}}{\partial t} = k_1 e'_{ij} + \mu\left(1 + \frac{k_1}{k_2}\right) \frac{\partial e'_{ij}}{\partial t}. \tag{5}$$

The coefficients k_1 and k_2 in Equation 5 represent elastic moduli measured in dynes/cm^2 and μ is a viscosity measured in dynes-sec/cm^2.

To utilize the model indicated in Equation 5 in any particular case, it is necessary to solve the equations of motion subject to the appropriate boundary conditions. Assuming the deformation takes place slowly and the Reynolds number of the flow involved is small as usual in the microcirculation, inertial effects may be neglected and the equations of motion then reduce to the equilibrium equations

$$\frac{\partial \sigma_{ij}}{\partial x_j} = 0. \tag{6}$$

In the case of the pipette experiment illustrated schematically in Figure 2, the appropriate boundary conditions are taken to be uniform pressure on each part of the sphere, namely, atmospheric pressure in the outer surface A_0, the pipette pressure on the interior surface A_1, and a pressure P_g for the surface in contact with the glass pipette on area A_g. The pressure P_g can be computed on the basis of the equilibrium of the cell as a whole. The force on the cell due to the pressures P_0 and P_1 is equal to the area of the lumen of the pipette times the pressure difference $(P_0 - P_1)$. The analytical problem of the deformation of a viscoelastic sphere under

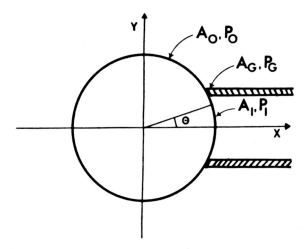

Fig. 2. Schematic of a leukocyte test by micropipette technique. The areas A_I, A_G, A_0 refer to the cell surface inside the pipette, the area of contact with the glass of the pipette, and external fluid area respectively. P_I, P_G, P_0 are the corresponding pressures on these surfaces.

micropipette suction as shown in Figure 2 has been solved in closed form for small strains by Schmid-Schönbein et al [1981]. The solution is time dependent and is proportional to the creep function $J(t)$ of the standard viscoelastic solid model indicated in Figure 1. This creep function is

$$ J(t) = \frac{1}{k_1} \left[1 - \left(1 - \frac{k_1}{k_1 + k_2} \right) \exp \left(- \frac{k_1 k_2 t}{\mu(k_1 + k_2)} \right) \right]. \tag{7} $$

A comparison of experimental and theoretical results in the case of aspiration of a leukocyte into a pipette is made on the basis of the distance $d(t)$ that the cell is drawn into the pipette. A computer program determines the coefficients k_1, k_2, and μ required for the optimum fit of the analytical solution to the experimental data. A typical result is shown in Figure 3. Summary results for a large number of tests under various experimental conditions are given in the following paper. In general, the micropipette experiments are well represented by the present model.

THEORY OF ACTIVE LEUKOCYTE DEFORMATION

The active formation of protopods of leukocytes has been observed both in free suspension in plasma and in cells adhering to a substrate [Schmid-

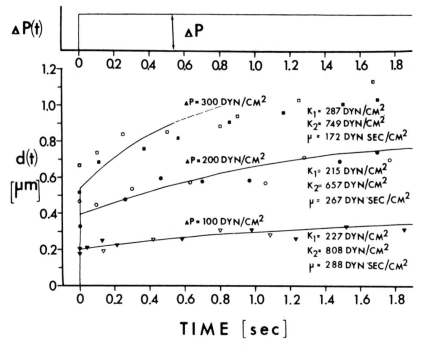

Fig. 3. The recorded pressure $\Delta P(t)$ applied to a leukocyte by micropipette and the displacement $d(t)$ into the pipette as functions of time. The different symbols represent different measurements on a single neutrophil suspended in a medium of 310 mOsm at 22°C. The solid curve is the theoretical result using best-fit coefficients k_1, k_2, and μ listed.

Schönbein et al, 1982]. The formation of protopods in the free suspension indicates that the energy and forces required can be of internal origin and not dependent on adhesive forces. Electron micrographs such as shown in Figure 4 indicate that the protopod cytoplasm has a uniform texture and is devoid of granules or nucleus of the cell. Micropipette tests indicate that the protopod is stiffer than the body of the cell. These observations and biochemical analyses have led to the conclusion that the protopods consist of a gel which is primarily a polymerized actin network.

A continuum theory based on the polymerization of the actin matrix in the cell as the driving agency of protopod formation has been proposed by Schmid-Schönbein and Skalak [1983]. It is postulated that the gelation is triggered by the influx of $Ca++$ across specialized regions of the cell membrane and that polymerization occurs on an interface at the base of the protopod where it connects to the cell. In this theory, the interface

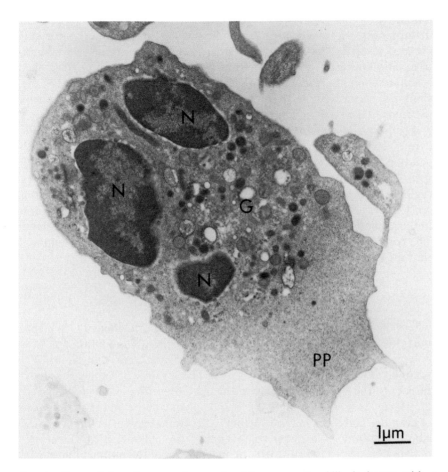

Fig. 4. Transmission electron micrograph of human neutrophilic leukocyte with a protopod. The cell was freely suspended in Ringer's solution at the time of chemical fixation. The cytoplasm in the protopod (PP) is devoid of granules (G) and other cell organelles, and the cell membrane is largely unfolded over the protopod.

moves as calcium diffuses through the protopod toward the cell body. The protopod grows outward rather than inward into the cell because of the balance of the forces acting. The external plasma medium is of low viscosity and offers little resistance. The interior of the cell is much more viscous and stiff, as discussed in the preceding section. The cell membrane, when not stretched taut, offers little resistance, and the balance of forces then favors the outward projection of the protopod with simultaneous unfolding of the membrane. In osmotically swollen cells where the membrane is stretched taut, it can be shown that polymerization does proceed inward. This counter example reinforces the proposed assumption. The notion of a gelation front also accounts for the exclusion of granules from the pseudopod. The process may be visualized as being similar to the advancing front of a frozen zone of a liquid. As such a front proceeds, it may exclude impurities as it advances into the liquid.

For purposes of analysis of active formation of pseudopods, the leukocyte is considered to consist of two different regions separated by the gelation front. The body of the cell will be designated as a region 1 and the pseudopod is considered a separate region 2. The main body of the cell in region 1 is considered to have the properties discussed in the preceding section and to be described by the Equations 1–7. The equations of motion in the pseudopod region 2 are the same as in region 1, ie, Equation 6 holds. The constitutive equation in region 2 is taken to be an elastic medium so that the Equation 5 is replaced by

$$\sigma_{ij}^{(2)} = 2G(e_{ij}^{(2)} - e_{ij}^{(0)}) - p^{(2)}\delta_{ij}, \tag{8}$$

where the superscript (2) indicates the region. The coefficient G is a shear modulus of the polymerized material. The strain $e_{ij}^{(0)}$ is a strain due to the polymerization induced by the gelation process and the exclusion of granules by the interface as it moves. The strain $e_{ij}^{(2)}$ is measured with respect to the state in the body of the cell before the material crosses the interface. As it does so, there is a shrinkage in volume due to the exclusion of granules and there may be further a directional polymerization strain at constant volume. The shear component of the polymerization strain $e_{ij}^{(0)}$ may be considered to be the driving mechanism of the protopod formation. It is enforced by the gelation process. When the strain in the polymerization region $e_{ij}^{(2)}$ is equal to $e_{ij}^{(0)}$ there are no stresses in the pseudopod other than the ambient pressure $p^{(2)}$ as indicated by Equation 8.

The polymerization strain field $e_{ij}^{(0)}$ is conveniently separated into two stages. The first is a volume change which is isotropic and is required by the exclusion of the granules. In terms of principal stretch ratios λ_i^v, the ratio of the final volume of an element dV to the initial volume dV_0 is

$$dV/dV_0 = \lambda_1^v \lambda_2^v \lambda_3^v = \theta. \tag{9}$$

By definition the isotropic volume change θ is such that λ_1^v, λ_2^v, λ_3^v are all equal. Any further change of shape at constant volume is described by a second component λ_i^P. In this case considering the direction of x_1 to be perpendicular to the gelation front, the two components λ_2^P, λ_3^P would be equal and

$$\lambda_1^P \lambda_2^P \lambda_3^P = 1. \tag{10}$$

Equation 10 indicates the preservation of volume during the polymerization strain. The cross sectional area of a filament as it passes through the gelation front is reduced in area by the ratio A_r given by

$$\lambda_2^P \lambda_3^P = A_r. \tag{11}$$

It can be shown that specifying the area ratio A_r and the volumetric ratio θ given by Equation 9 determines the λ_i^v, λ_i^P [Schmid-Schöenbein and Skalak, 1983]. To complete the theory of the protopod formation, it is necessary to add a law for the movement of the gelation front and to determine how the various parts of the cell will move in response to the advancing gelation. It is assumed that increasing local calcium concentration at the gelation front causes the polymerization of actin at a certain critical value. In order to produce a solation, some antagonist must be invoked to reverse the process. The diffusion of calcium is assumed to proceed from the cell membrane toward the body of the cell through the protopod and to be governed by the diffusion equation

$$\frac{\partial c}{\partial t} = \nabla \cdot (D_c \nabla c) + \mathbf{v} \cdot \nabla c, \tag{12}$$

where ∇ is the gradient operator, $\mathbf{v}$ is the cytoplasm velocity, D_c is the diffusion coefficient for calcium, and c is the calcium concentration at any point. At the interface, it is assumed that when actin polymerizes, it absorbs an amount of calcium A_c moles per unit of actin volume polymerized. The calcium required must be provided by a difference in the fluxes of calcium on the two sides of the interface. Thus

$$[(D_c \nabla c)^+ - (D_c \nabla c)^-]\Delta t = A_c \Delta x, \tag{13}$$

where $D_c \nabla c$ represents the local calcium flux and the superscripts $\pm$ indicate the two sides of the interface and Δx is the displacement of the interface during a small time interval Δt. The velocity of the interface is $v_I = \Delta x / \Delta t$, which may be expressed as

$$v_I = [(D_c \nabla c)^+ - (D_c \nabla c)^-]/A_c. \tag{14}$$

It follows from the assumptions that the difference in calcium concentration across the interface is

$$c^+ - c^- = A_c. \tag{15}$$

Equations 12, 14, and 15 govern the flux of calcium and thereby the velocity of motion of the gelation interface.

The interface velocity given by Equation 14 is relative to the gelled region of the pseudopod. The relative motion of the pseudopod with respect to the cell body requires an additional consideration of the forces involved, as shown in Figure 5. The position of the interface at the base of the pseudopod is shown at a time t_0 in Figure 5A. Figure 5B, C shows two different possibilities at a time Δt later when the wavefront has advanced by an amount Δb. In Figure 5B the protopod has moved outward so the interface remains at the surface of the body of the cell. In Figure 5C there has been no relative motion of the protopod and the cell body, so that the motion of the interface results in the intrusion of the pseudopod into the cell body by the amount of Δb. In order to distinguish which possibility arises in fact, the force system acting at the interface is considered in Figure 5D. If the overall length L of the cell and protopod increase, there must be some displacement of the external fluid which will result in the fluid drag forces F_1 and F_2 shown in Figure 5D. On the interface the force between the cytoplasm of the cell and the gelated material in the protopod is indicated as an internal force F_i. The force in the membrane, if any, is designated as the tension force F_T. For equilibrium of each part of the cell, it is required that

$$F_i = F_1 + F_T = F_2 + F_T. \tag{16}$$

To estimate each of the forces involved, consider first that the total length of the cell L may be written as

$$L = b_1 + b_2 - \xi, \tag{17}$$

where b_1 is the diameter of the body of the cell, b_2 is the length of the protopod, and ξ is the depth of penetration of the gelated region into the cell body. The fluid drag forces F_1, F_2 may be estimated as a Stokes drag on a sphere whose volume is one-half of the total cell volume. The relative velocity of the two parts of the cell is L. The velocity of each part of the cell

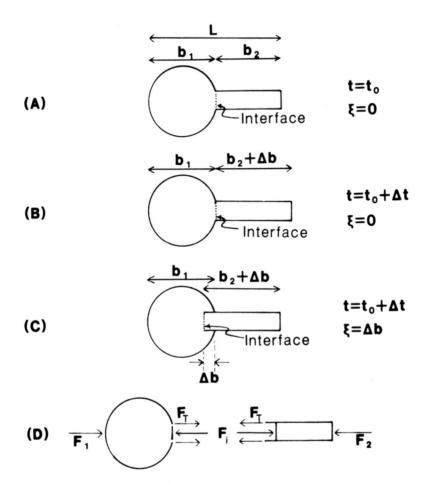

Fig. 5.   Schematic of protopod motion. A. The total length of the cell L is the sum of the cell body diameter b_1 and the protopod length b_2. B. When b_2 increases, the overall length may increase if the interface remains at the surface of the cell body or (C) the overall length could remain fixed if the protopod grows into the cell body by the penetration depth $\xi = \Delta b$. D. The force system acting at the interface. F_1 and F_2 are fluid drag forces. F_T is due to membrane tensions and F_i is the internal stress at the interface (from Schmid-Schönbein and Skalak [1983]).

may be taken as $\dot{L}/2$, so that the Stokes drag forces become

$$F_1 = F_2 = 6\pi\mu \frac{b_1}{2} \frac{\dot{L}}{2} = K_1\dot{L}, \tag{18}$$

where the super dot indicates differentiation with respect to time and μ is the viscosity of the external fluid. The force due to membrane tension may be estimated by the form

$$F_T = (T_0 + K_2(b_2 - b_{20} - \xi))S, \tag{19}$$

where T_0 is the tension in the membrane when b_2 is equal to b_{20} and K_2 is a modulus of stiffness of the membrane. The length S in Equation 19 is the perimeter of the cross section of the membrane on which this tension acts.

The internal force F_i will depend primarily on the penetration ξ of the pseudopod into the cell body. Assuming a viscoelastic behavior of the cell body, the form of F_i is taken to be

$$F_i = K_3\xi + K_4\dot{\xi}, \tag{20}$$

where K_3 and K_4 are coefficients to be determined from the passive properties of the cell body. Substituting the estimates (Eqs. 18–20) into the equilibrium Equation 16 yields the equation

$$K_3\xi + K_4\dot{\xi} = K_1(\dot{b}_2 - \dot{\xi}) + (T_0 + K_2(b_2 - b_{20} - \xi))S. \tag{21}$$

In Equation 21, the value of $\dot{L}$ has been replaced by

$$\dot{L} = \dot{b}_2 - \dot{\xi}. \tag{22}$$

Equation 21 may be regarded as a differential equation on ξ with the term b_2 as the driving force of the equation. The value of $\dot{b}_2$ is the relative rate of advance of the gelation front given by Equation 14. It is expected that the constants K_1, T_0, K_2S in Equation 21 will be much smaller than the constants K_3 and K_4. In this case, the solutions of Equation 21 will yield ξ approximately zero for all times and the pseudopod will grow outward as indicated in Figure 5B. In a swollen cell the opposite occurs. The membrane term K_2 dominates Equation 21, which then has the solution $\xi = b_2$ and $\dot{L}$ is then zero. In this case, the full rate of increase of gelated length b_2 is absorbed inside the cell with the overall length remaining constant as shown in Figure 5C.

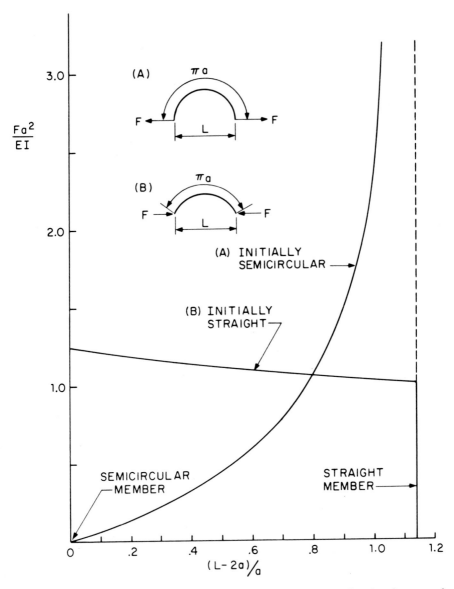

Fig. 6.   The forces required to bend or unbend a membrane. The length and surface area of the membrane are assumed to be fixed, and only bending stiffness is taken into account. Curve A shows the force required to pull a semicircular membrane toward a flat. Curve B shows the force required to buckle and bend a flat membrane into a buckled (curved) shape. E is the Young's modulus and I the area moment of inertia of the membrane cross section.

To estimate the constants in Equation 21, the following approximations are made. The viscosity of the external fluid is taken to be 1 centipoise (cp), and the initial diameter of the cell b_1 is taken to be 7.9 μm. To estimate the constants K_3 and K_4 the indentation of the cell is approximated by a state of plane stress of an incompressible material in which

$$\sigma_{ij} = \begin{bmatrix} \sigma_x & 0 & 0 \\ 0 & 0 & 0 \\ 0 & 0 & 0 \end{bmatrix} \qquad e_{ij} = \begin{bmatrix} e_x & 0 & 0 \\ 0 & -e_x/2 & 0 \\ 0 & 0 & -e_x/2 \end{bmatrix}. \qquad (23)$$

Then the governing viscoelastic equation (Eq. 5) can be reduced to

$$\sigma_x = \frac{3}{2} e_x k_1 + \frac{3}{2} \dot{e}_x \mu. \qquad (24)$$

In deriving Equation 24, it has been assumed that k_2 is much larger than k_1. (Average values are $k_1 = 275$ dynes/cm^2, $k_2 = 737$ dynes/cm^2, $\mu = 130$ dynes-sec/cm^2.) Using these values for k_1 and μ and assuming the cell to be replaced by a cylindrical shape 4 μm in length and 4 μm leads to numerical estimates of K_3 and K_4. The cylinder used here is intended to represent the part of cell which is most stressed.

To estimate the constants T_0 and K_2 in Equation 21, an analysis is made of the force required to unbend a semicircular cylinder of membrane shown in Figure 6. These results have been computed on the basis of elastic theory [Timoshenko and Gere, 1961]. The tension per unit length starting from a semicircle of radius a is approximated by

$$T = 1.32 \frac{B}{a^2} \epsilon, \qquad (25)$$

where $\epsilon = \Delta L/L$ is the strain based on the initial semicircle diameter, and B is the bending modulus. The value of B for white cell membrane is not known at this time, but is here taken to be $B = 10^{-12}$ dyne cm, which is the value for red blood cell membrane. The length S in Equation 21 is taken to be the circumference of a pseudopod 2 μm in diameter. With all of the above numerical estimates, Equation 21 can be written in the following form:

$$0.130\xi + 6.13 \times 10^{-2} \dot{\xi} = 3.74 \times 10^{-5}(b_2 - \dot{\xi}) + 7.79$$

$$\times 10^{-8} + 2.51 \times 10^{-4}(b_2 - b_{20} - \xi), \qquad (26)$$

where each term is a force in dynes when ξ, b_2, b_{20} are in cm and $\dot{\xi}$ and $\dot{b}_2$ are in cm/sec. It can be seen from the magnitudes of the constants in Equation 26 that solutions of the equation for ξ will generally give a small fraction of a micron for small changes in b_2. In the above estimates for membrane tension, it is assumed that the initial shape of a typical wrinkle in the membrane is a semicircular form. If the initial shape is flat, so that the membrane is being forcibly distorted in the normal wrinkled shape, the elastic energy of the membrane would be helping to unfold it during the pseudopod formation. In this case, the stress in the membrane would be effectively compression, which would help to drive the pseudopod outward. The results, shown in Figure 6, indicate that in either case the resultant forces are small and negligible compared to the stresses that can be developed in the cytoplasm.

In the analyses leading to the results in Figure 6, the membrane shear elasticity has been neglected because the wrinkles of the leukocyte are often in ridges which approximate a two-dimensional form. Such a membrane is applicable to a flat surface so no membrane stresses due to shear in the plane of the membrane would occur in the unfolding of such a wrinkle. These factors also support the neglect of membrane stresses in analyzing the micropipette tests of leukocytes in the small deflection range.

CONCLUSIONS

The models of passive and active deformation of leukocytes proposed in the preceding sections are in qualitative agreement with many aspects of experimental data. The passive properties for small deformations are more firmly established numerically by micropipette tests at this time. The analyses show that the viscoelastic behavior of leukocytes must be primarily attributed to the cytoplasm. Membrane stresses are negligible until the membrane is stretched taut.

The model of active deformation and formation of pseudopods suggested above is driven by the polymerization process which produces a gel network primarily of actin. This model is compatible with known characteristics of pseudopods, but more detailed verification of the assumptions is needed to establish the mechanisms definitively. The theory also will require refinement such as treating the nucleus of the cell as a separate body and considering the cytoplasm as a two or more phase composite such as granules, actin, myosin, and microtubules. When the properties of each of these constituents is better known and their geometric and chemical interaction is established, a more complete model can be generated.

ACKNOWLEDGMENTS

The authors are grateful to Paul Lui, who assisted in the computations. This research was partially supported by NIH grant HL-16851.

REFERENCES

Bagge U, Brånemark P-I (1977a): White blood cell rheology. An intravital study in man. Adv Microcirc 7:1–17.
Bagge U, Skalak R, Attefors R (1977b): Granulocyte rheology. Adv Microcirc 7:29–48.
Branemark P-I (1971): Intravascular Anatomy of Blood Cells in Man. Basel: S. Karger, pp 43–55.
Crank J (1967): The Mathematics of Diffusion. London: Oxford University Press.
Dembo M, Tuckerman L, Goad W (1981): Motion of polymorphonuclear leukocytes: Theory of receptor redistribution and the frictional force on a moving cell. Cell Motil 1:205–235.
Fung YC (1965): Foundations of Solid Mechanics. Englewood Cliffs, New Jersey: Prentice-Hall, Inc.
LaCelle PL, Bush RW, Smith BD (1982): Viscoelastic properties of normal and pathologic human granulocytes and lymphocytes. In Bagge U, Born GVR, Gaehtgens P (eds): "White Blood Cell. Morphology and Rheology as Related to Function." The Hague: Martinus Nijhoff Publishers.
Schmid-Schönbein GW, Sung KLP, Tözeren H, Skalak R, Chien S (1981): Passive mechanical properties of human leukocytes. Biophys J 36:243–256.
Schmid-Schönbein GW, Skalak R, Sung KLP, Chien S (1982): Human leukocytes in the active state. In Bagge U, Born GVR, Gaehtgens P (eds): "White Blood Cells Morphology and Rheology as Related to Function." The Hague: Martinus Nijhoff, pp 21–31.
Schmid-Schönbein GW, Skalak R (1983): Continuum mechanical model of leukocytes during protopod formation. J Biomech Eng (in press).
Skalak R, Schmid-Schönbein GW, Chien S (1982): Analysis of white blood cell deformation. In Bagge U, Born GVR, Gaehtgens P (eds): "White Blood Cells Morphology and Rheology as Related to Function." The Hague: Martinus Nijhoff, pp 1–10.
Stossel TP (1982): The structure of the cortical cytoplasm. Philos Trans R Soc Lond [Biol] 299:275–289.
Stossel TP, Pollard TD (1973): Myosin in polymorphonuclear leukocytes. J Biol Chem 248:8288–8294.
Timoshenko SP, Gere JM (1961): Theory of Elastic Stability. New York: McGraw-Hill, pp 76–82.

White Cell Mechanics: Basic Science and
Clinical Aspects, pages 19–51
© 1984 Alan R. Liss, Inc., 150 Fifth Avenue, New York, NY 10011

Viscoelastic Properties of Leukocytes

S. Chien, G. W. Schmid-Schönbein, K.-L. P. Sung, E. A. Schmalzer, and R. Skalak

Departments of Physiology, College of Physicians & Surgeons (S.C., K.-L.P.S., E.A.S.) and Civil Engineering and Engineering Mechanics, Bioengineering Institute (R.S.), Columbia University, New York, New York 10032 and Department of Applied Mechanics and Engineering Sciences, University of California at San Diego, La Jolla, California 92093 (G.W.S.-S.)

INTRODUCTION

The rheological behavior of blood cells plays an important role in governing blood flow dynamics in the microcirculation [1]. The white blood cells (WBCs) exist in a much lower volume concentration than the red blood cells (RBCs). Because of their larger volume and lower deformability [2, 3], however, the WBCs may exert significant rheological influences on microcirculatory flow [4, 5]. Knowledge on the rheological properties of WBCs is essential not only for the elucidation of microcirculatory flow dynamics, but also for the understanding of their other functional behaviors in health and disease.

The analysis of the rheological behavior of WBCs requires quantitative knowledge on their morphological characteristics, including the volume and membrane area of the cells and of their nuclei. Although the WBCs are spherical in outline, the presence of membrane foldings provides an excess membrane area which allows cell deformation without area expansion. These structural parameters, together with the mechanical properties of the cell components, determine the deformability of the WBCs, especially during large deformations, such as those involved in their release from the bone marrow or extravasation into the interstitium.

The WBCs are usually spherical in the arterioles and venules; they may be deformed when passing through the narrow capillaries, but they do not undergo spontaneous, active deformation during flow through the microcirculation [6, 7]. Spontaneous active deformation of WBCs with protopod formation occurs in vivo during extravasation across the endothelial wall [8], migration in a chemotactic gradient [9–11], and phagocytosis [12]. When human WBCs are kept in vitro in plasma or Ringer's solution, with the presence of free Ca^{2+} ions, they will undergo spontaneous deformation and project protopods without the action of an external mechanical force. If the WBCs are suspended in a medium with ethylenediamine tetraacetate (EDTA) present to chelate Ca^{2+}, the spontaneous, active deformation is eliminated and the WBCs remain essentially spherical for many hours [13]. We studied the mechanical properties of the human neutrophils in both the passive state (in the presence of EDTA) and the active state (in the presence of Ca^{2+}) by measuring their deformation in response to micropipette aspiration. These experiments were performed under small deformations so that the geometric factors are not important. The data were analyzed by using a theoretical model of a viscoelastic solid. Large deformations of WBCs were also investigated by studying their filterability through polycarbonate sieves with 5-μm pores.

This presentation serves to summarize several recent studies performed in our laboratory related to WBC rheology [13–17]. The aims of these investigations are to establish the mechanical properties of WBCs and elucidate the rheological basis of WBC functions in health and disease.

MATERIALS AND METHODS
Sample Preparation

Fresh blood samples were obtained from healthy human subjects using EDTA as anticoagulant. The red blood cells (RBCs) were allowed to sediment at room temperature for 25–40 minutes. The supernatant plasma containing WBCs, platelets, and a few RBCs was collected with a Pasteur pipette. This WBC-rich plasma was diluted with a prefiltered, buffered saline-albumin solution to a concentration of about 50 WBCs/mm^3. The solution contained 0.9 g/dl NaCl, 0.1 g/dl EDTA, 0.25 g/dl bovine serum albumin, and 12 mM Tris, with pH adjusted to 7.4 by the dropwise addition of 1 N HCl. The osmolality as measured by means of the freezing-point depression method was 300–310 mOsm/kg H_2O. In experiments in which the osmolality was altered, this was achieved by changing the concentration of NaCl. Alterations in pH were attained by changing the amount of Tris or HCl. In studies on neutrophil deformation in the active state, neutrophils obtained from fresh heparinized blood samples were suspended in a Tris-buffered Ringer's solution (pH 7.4) containing 2 mM Ca^{2+}, 0.25 g/dl bovine serum albumin, and no EDTA.

Electron Microscopy

The cells were fixed by adding a buffered 2% glutaraldehyde solution (pH = 7.4) into the cell suspension at a rate of about 1 drop per 10 sec, with continuous stirring; the final glutaraldehyde concentration was 1%. The cells were then fixed for an additional hour in a fresh 2% glutaraldehyde solution, washed in NaCl solution, postfixed with 1% OsO_4 for 1 hour, and washed in cacodylate buffer and in distilled water. These postfixed cells were then processed for scanning electron microscopy (SEM) and transmission electron microscopy (TEM).

For SEM, the postfixed cells were dehydrated in ascending series of ethanol and either critical-point or air dried. The specimens were glued to aluminum stubs with silver paint, coated with gold palladium using a Hummer I sputterer, and examined and photographed in a Jeolco JSM-25 scanning electron microscope at 15 kV.

For TEM, the postfixed cells were stained in tannic acid [18], rinsed in distilled water, and stained with 2% uranyl acetate. After rinsing in distilled water, the cells were dehydrated in ascending series of ethanol, rinsed three times in propylene oxide, infiltrated with 50% Epon in propylene oxide for 24 hours, embedded in 100% Epon, and hardened at 60°C. The embedded specimen was sectioned with a diamond knife on an automatic microtome (Dupont/Survall MT2-B) at approximately 60–80-nm thickness (silver section). The sections were carried on a copper mesh, stained with 6% uranyl magnesium acetate and lead Reynold citrate, and examined and photographed on a Zeiss (EM 9) transmission electron microscope.

Morphometry

The three-dimensional geometry of individual WBCs was calculated with the aid of stereology [19, 20] from TEM thin sections at random radial positions of the WBCs. All WBCs studied show an overall spherical shape, and in isotonic or hypertonic solutions they have numerous membrane foldings. All random TEM cell sections were separated into four groups: neutrophils, lymphocytes, monocytes, and eosinophils (the number of basophils were insufficient). On each section a quadratic test line system with grid size 0.3 or 0.4 cm was superimposed for morphometric measurements.

The WBCs are considered to be a polydispersed system of spheres contained in a control volume. The sphere diameters D_j are grouped in K intervals with step size Δ so that $D_1 = 1\Delta$, $D_2 = 2\Delta$, until $D_k = D_{max} = k\Delta$. There are $(N_v)_j$ WBCs per unit control volume with sphere diameter D_j. On random sections the WBCs yield section diameter d_i, with $(N_A)_i$

number of cell sections per section area through the control volume. d_i and $(N_A)_i$ are measured from the TEM micrographs with an areal count [11] and the d_i values are also grouped in K intervals with step size Δ. $(N_v)_j$ is then computed according to the equation

$$(N_v)_j = \frac{1}{\Delta} [\alpha_j(N_A)_j - \alpha_{j+1}(N_A)_{j+1} - \ldots - \alpha_k(N_A)_k]. \tag{1}$$

The average cell volume V_c is then

$$V_c = \frac{\sum\limits_{j=1}^{k} (N_v)_j \, D_j^3 \, \frac{\pi}{6}}{\sum\limits_{j=1}^{k} (N_v)_j}. \tag{2}$$

The average cell membrane area (S_c) in a population of WBCs with average volume V_c can be measured from random sections as

$$S_c = V_c \times \frac{2P_L}{P_p} \bigg|_{Cell,} \tag{3}$$

where P_L is the sum of all intersections between the test line system and the cell membrane divided by the length of the test lines, and P_p is the number of test points inside all cell sections divided by the number of test points on the grid. The same relation can be applied to derive the surface/volume ratio of the cell nucleus.

Micropipette Aspiration

The experimental set up and procedure have been described elsewhere [13, 21]. About 0.5–1-ml of the cell suspension was loaded in a small round chamber located on the stage of an inverted microscope (Fig. 1). The chamber was surrounded by a copper ring filled with methanol-water (1:3), which was circulated from a temperature control reservoir. A thermistor probe was present in the sample chamber to monitor the temperature of the cell suspension. Individual WBCs were viewed with an 100× objective (NA 1.25, oil immersion) and a 20× eyepiece. The viewing field was displayed on a video monitor through a video camera connected to the eyepiece, and the video image was recorded on a video recorder.

Micropipettes with internal radii of 1.1–1.7 μm were filled with the buffered saline-albumin solution and mounted on a hydraulic micromanipulator. The wide end of the micropipette was connected to a pressure

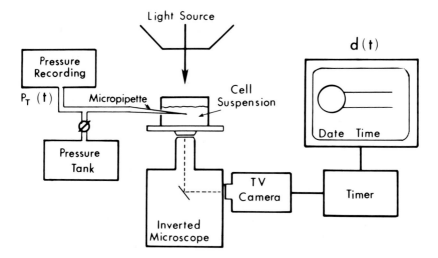

Fig. 1. A schematic drawing of the system used for the micropipette experiment (from [15]).

regulation system. The pressure level was monitored with the use of a Statham transducer connected to a Gould recorder.

To analyze the time course of deformation of the neutrophils, sequential photographs were taken from the video image during single-frame replay on the video monitor. The displacement of the cell surface into the pipette was determined by subtracting the distance that the cell reaches into the pipette without deformation. The displacement data were entered into a PDP 11/10 minicomputer and analyzed by using our theoretical model [13], in which the neutrophil is treated as a standard solid consisting of an elastic element K_1 in parallel with a Maxwell element (an elastic element K_2 in series with a viscous element μ).

Microsieving of Leukocytes Through Polycarbonate Filters

The filterability of WBC suspensions through polycarbonate sieves were studied by the constant flow technique described by Usami et al [22, 23]. Suspensions of RBCs and WBCs in various proportions were prepared from fresh human blood samples, using a Tris-buffered Ringer's solution (pH 7.4, containing 0.5 g/dl serum albumin) as the suspending medium. The suspension was pumped at a constant flow rate (Q) of 1.6 ml/min through 13-mm polycarbonate sieves (Nuclepore Corp., Pleasanton, CA) with a mean pore diameter of 4.79 ± 0.19 (SD) μm, a pore density of $3.6 \times 10^5/cm^2$ and an effective filtration area of 0.79 cm^2. The filtration pressure (P) was recorded as a function of time (t) with the use of a Statham pressure transducer and a Grass recorder. Filtration pressure (P_0) was also determined for the Ringer solution. In some of the experiments, at the end of the test, the filter was removed and processed for SEM and TEM examinations.

RESULTS

Morphometric Data

Morphometric measurements were made on WBCs in EDTA. All WBCs studied were spherical in their outline and there was no protopod formation. The TEM micrographs in Figure 2 show sections of examples of neutrophils, lymphocytes, monocytes, and eosinophils, fixed in media at 310 mOsm (left) and 105 mOsm (right). The appearance of the cells in hypertonic solution is similar to that in isotonic medium in that they all have a folded membrane, whereas in hypotonic solution the membrane becomes unfolded and the cell approximates the shape of a smooth sphere.

Table I shows the average volumes and membrane areas of various types of WBCs and their nuclei in isotonic medium. The percent excess membrane area σ was calculated from the difference between the average cell membrane area (S_c) and the area needed to cover its volume V_c by a sphere (S_s).

$$\sigma = 100(S_c - S_s)/S_s. \qquad (4)$$

It is to be noted that the spherical WBCs, because of the presence of membrane foldings, have even more excess membrane area than the discoid RBCs. The value of σ varies among the various types of WBCs studied. The nuclei of the WBCs also have excess membrane areas. In an isotonic medium, the average excess membrane areas varied from 84% to 137% for different types of WBCs and from 40% to 97% for their nuclei.

When suspended in hypotonic media, the cell and the nucleus increase their volumes while their membrane areas remain essentially constant, and the values of σ decrease. When the cell or its nucleus is swollen to a sphere with a smooth surface, then $\sigma = 0$. For neutrophils, the nucleus and the cell approach spherical shape at comparable levels of hypotonicity. For lymphocytes, however, their nuclei become spherical when their cell membrane is still partially folded.

The fractional volume content of the cell nucleus, granules, mitochondria and the remainder, collectively designated as cytoplasm, are listed in Table II for WBCs in an isotonic medium. The cytoplasm contains the Golgi region, the rough endoplasmic reticulum, and the remaining cell matrix, which occupies a large portion of the cell volume.

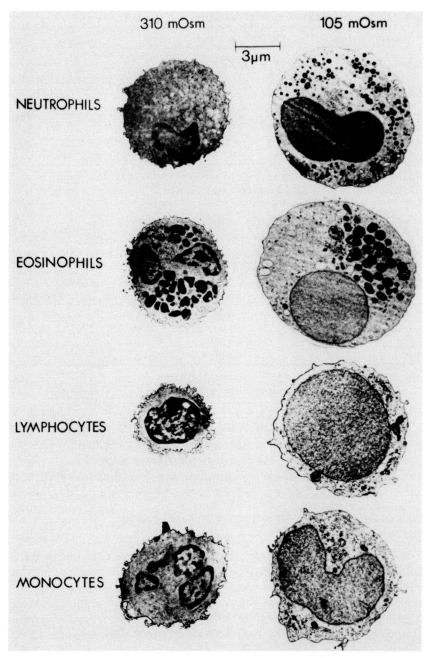

Fig. 2. Thin sections through human neutrophils, eosinophils, lymphocytes, and mono-cytes at 310 mOsm (left) and 105 mOsm (right). The length of the crossbar corresponds to 3 μm. Note that the radial position at which the section was made is unknown (from [14]).

TABLE I. Geometric Characteristics of Human Erythrocytes and Leukocytes

	Whole Cell				Nucleus		
	Volume (μm^3)	Surface area		$D_{min}{}^a$ (μm)	Volume (μm^3)	Surface area	
		μm^2	Excess			μm^2	Excess
Erythrocytes	90	140	44	2.7	—	—	—
Neutrophils	190	300	84	2.6	35	102	97
Eosinophils	206	324	92	2.6	37	99	84
Lymphocytes	120	270	130	1.8	50	90	40
Monocytes	230	430	137	2.2	52	108	62

[a]Minimum diameter the cell can achieve at constant area and volume for a cylindrical shape with hemispherical caps; the influence of nucleus on D_{min} is ignored in the calculation.

TABLE II. Leukocyte Volume Distribution[a]

	Percent of cell volume			
	Nucleus	Granules	Mitochondria	Cytoplasm
Lymphocyte	44.4	0.2	3.8	51.6
Neutrophil	21.3	15.4	0.6	62.7
Monocyte	25.9	2.6	3.5	68.1
Eosinophil	17.9	23.0	0.7	58.4

[a]Determined at 310 mOsm and pH = 7.4 (from [14]).

The membrane foldings of WBCs can be seen on scanning electron microscopy of cells fixed in suspension (Fig. 3). Scanning electron microscopic examination of WBCs dried after preparation of a smear on a glass slide shows that the WBCs are flattened to a pancake shape with diameters larger than 10 μm (Fig. 4).

Passive Deformation of Neutrophils in Response to Micropipette Aspiration

When WBCs are suspended in EDTA, which causes chelation of Ca^{2+}, there was no active, spontaneous deformation of the WBC during the time period (<4 hours) elapsed between sampling and measurement. Therefore, the passive mechanical behavior of the WBC was studied in these experiments. The time course of deformation of a neutrophil in saline-albumin solution (pH 7.4, 300 mOsm, 21–23°C) in response to an aspiration pressure (ΔP = 3 mm H_2O) applied via a micropipette (radius = 1.7 μm) is illustrated by the example in Figure 5. The WBC was rapidly

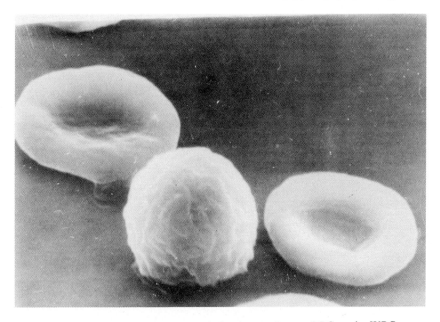

Fig. 3. Scanning electron micrograph showing two human RBCs and a WBC.

displaced toward the pipette tip and sealed it within one or two television frames (16–32 msec). The first frame in which the WBC is seen to make contact with the pipette tip is denoted zero time (0 msec in Fig. 5); at this moment the WBC already shows some deformation. The initial, rapid phase of deformation was essentially synchronous with the applied pressure (Fig. 6). This elastic response was followed by a creep displacement that was nonlinear with time. The dashed line for d(t) in Figure 6 represents the best fit of the standard solid model [7] to the experimental data, and the coefficients K_1, K_2, and μ are listed in the figure. Similar results were obtained for other neutrophils and for peripheral B lymphocytes. The peripheral T lymphocytes tend to have higher viscoelastic coefficients. The viscoelastic coefficients obtained on neutrophils and lymphocytes suspended in isotonic saline-albumin solution (pH 7.4, 310 mOsm, 21–23°C) are listed in Table III.

In general, the model fits the data well when d(t) is less than approximately 0.9 μm, but for larger deformations the model overestimates the experimental results. When different parts of the same WBCs were tested by the application of the same aspiration pressure, the d(t) curves are in close agreement, indicating that the cell has relatively homogeneous viscoelastic properties.

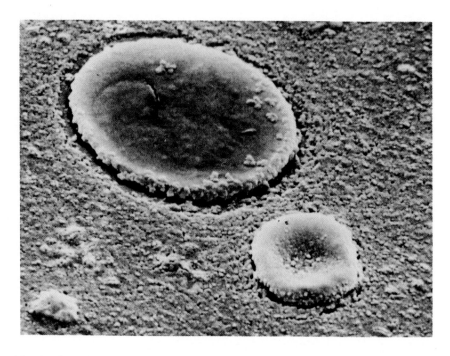

Fig. 4. Scanning electron micrograph of human WBC and RBC on a blood smear. The WBC is deformed into pancake shape with a diameter greater than 10 μm.

To test the assumption of linearity as well as the validity of the model, we applied, after the first pressure step at t = 0, a second step at t > 0 with an additional aspiration pressure ΔP. Figure 7 shows the results of one such experiment on a neutrophil. The coefficients K_1, K_2, and μ were computed only for the first step (for times t < 0.5 sec). The theoretical line drawn for the second step represents a prediction based on the coefficients derived from the first step. This prediction agrees well with the experimental data obtained from the second step. Similar agreement was obtained when the first step of aspiration pressure was followed by a step decrease in the pressure level (Fig. 8). The assumption of linearity of the system is further validated by experiments in which the same WBC was subjected to three separate aspirations at different pressure levels. As long as the displacement is less than 0.9 μm, the values of d(t) are proportional to the ΔP level, resulting in essentially the same viscoelastic coefficients at different ΔP.

The passive elastic behaviors of neutrophils and RBCs can be compared by plotting their steady-state deformation (D_{pm}) normalized for the pipette radius (R_p) against the stress parameter (ΔP)R_p (Fig. 9). Using micropi-

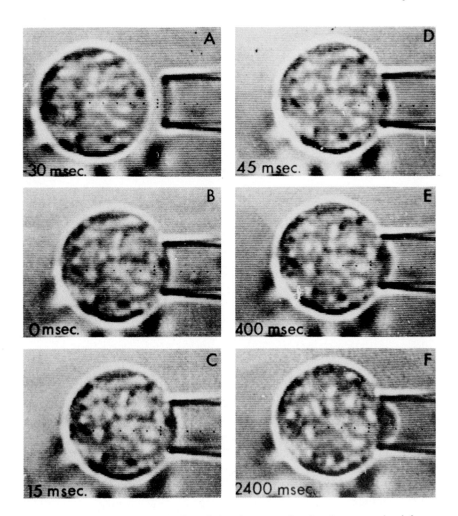

Fig. 5. Sequence of photographs from light microscopy showing the progressive deformation of a neutrophil at the indicated times after the cell had made contact with the pipette. Applied pressure $\Delta P = 500$ dynes/cm^2. The cell was suspended in a medium of 310 mOsm at 22°C and pH 7.4 (from [13]).

pettes with $R_p \simeq 0.5$ μm, in order to get a comparable D_{pm}/R_p, the value of $(\Delta P)R_p$ needed for neutrophils was approximately four times greater than that for RBCs. The time constant of the neutrophil deformation can be calculated as

$$\tau = \frac{\mu(K_1 + K_2)}{K_1 K_2} . \tag{5}$$

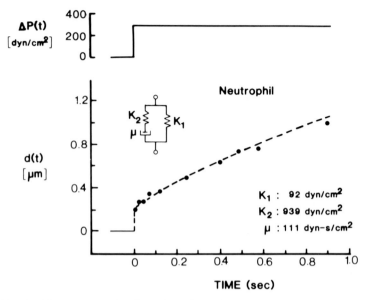

Fig. 6. Time courses of deformation, d(t), of a human neutrophil in EDTA-saline in response to a step aspiration pressure, ΔP(t). The data points were fitted with the theoretical model (dashed line) by the use of the K_1, K_2, and μ values tabulated in the figure.

The value was approximately 650 msec. The deformation of RBC membrane shows a rapid deformation phase with a time constant ranging from 20 to 120 msec, which varies inversely with the rate of deformation [21]. The viscosity of the neutrophil interior is approximately 130 P. In comparison, the viscosity of the normal RBC interior is only approximately 0.07 P [24, 25]. The rheological properties of human neutrophils and erythrocytes are compared in Table IV.

Effects of Physicochemical Factors on Rheological Properties of Neutrophils

The effects of variations in temperature, pH, and osmolality, as well as the addition of colchicine, on the passive rheological properties of neutrophils in EDTA were studied.

Temperature. The effects of temperature variation (range 9°–40°C) were studied in 112 experiments on 26 neutrophils. As shown in Figure 10, the elastic behavior (K_1 and K_2) of the neutrophils did not vary significantly with temperature. The μ value of the neutrophils varied inversely with temperature. The speed of Brownian motion of granules in the neutrophils increased with rising temperature, as μ decreased.

TABLE III. Viscoelastic Coefficients of Leukocytes in Small Passive Deformation[a]

Coefficients	Neutrophils	B Lymphocytes	T Lymphocytes
K_1 (dynes/cm^2)	275 ± 14	221	451
K_2 (dynes/cm^2)	737 ± 40	502	819
μ (dynes-sec/cm^2)	130 ± 6	145	206

[a]The leukocytes were suspended in isotonic saline-albumin-EDTA solution (pH 7.4, 310 mOsm, 21–23°C). Values given for neutrophils (n = 75) are mean ± SEM.

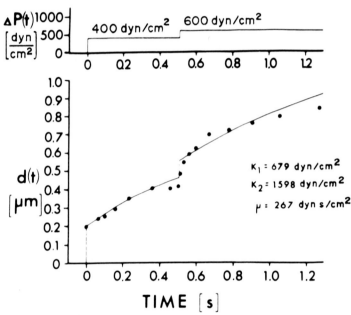

Fig. 7. The aspiration pressure ΔP and displacement d(t) for a two-step loading experiment of a human neutrophil in EDTA-saline. The coefficients were computed only from the data during the first-step loading. The line drawn for the second-step loading represents predictions by using the values of K_1, K_2, and μ obtained during the first step, (from [13]).

pH. The effects of pH variations (5.4–8.4) were determined in 115 experiments on 21 neutrophils. K_2 remained essentially constant in different pH media, but K_1 and μ increased with increasing pH (Fig. 11). In the higher pH media (pH 7.8 and 8.4), the cell diameter increased, and the granules also became swollen. In the lower pH media (5.4 and 6.0), the cells did not change their size, but they adhered more easily to the pipette tip and the bottom of the chamber and tended to form protopods.

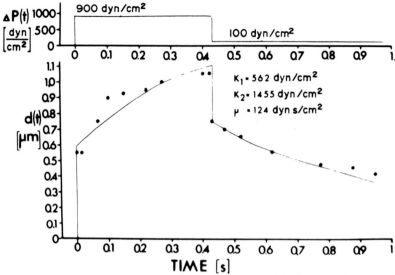

Fig. 8. The aspiration pressure ΔP and displacement d(t) for a loading-unloading experiment of a human neutrophil in EDTA-saline. The coefficients were computed only from the data during the loading (deformation) phase. The line drawn for the unloading (recovery) phase represents predictions based on the values of K_1, K_2 and μ obtained during the loading phase (from [13]).

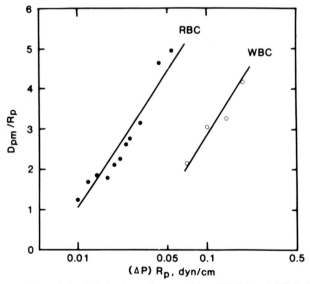

Fig. 9. Comparison of elastic behaviors of neutrophils (WBCs) and RBCs in response to an aspiration pressure ΔP applied via a micropipette with an internal radius R_p. To obtain a given degree of the elastic response, as represented by the steady-state deformation (D_{pm}) normalized for the pipette radius, the deforming stress required for WBCs is approximately four times that for RBCs.

TABLE IV. Comparison of Rheological Properties of Human Neutrophils and Erythrocytes

	Erythrocytes	Neutrophils
$(\Delta P)R_p$ needed for $D_{pm}/R_p = 3$ (dynes/cm)	0.025[a]	0.10
Time constant for small deformation (msec)	20–120[a,b]	650
Cellular viscosity (poise)	0.7	130

[a]Reflecting membrane behavior.
[b]Time constant for initial, rapid deformation phase.

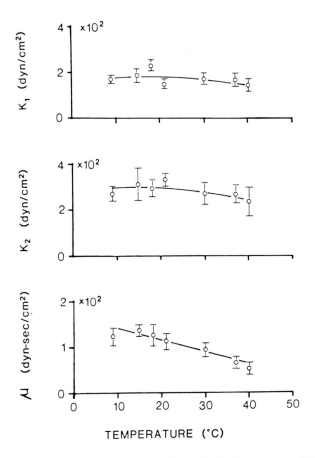

Fig. 10. The viscoelastic coefficients K_1, K_2, and μ for human neutrophils as a function of temperature. The cells were suspended in isotonic medium (310 mOsm) and at pH 7.4. Vertical bars represent standard errors of the mean.

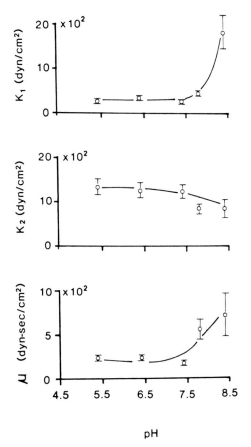

Fig. 11.   The viscoelastic coefficients K_1, K_2 and μ for human neutrophils as a function of pH. The cells were suspended in isotonic medium (310 mOsm) at room temperature (21–23°C). Vertical bars represent standard errors of the mean.

Osmolality. The effects of variations in osmolality (52–664 mOsm) were studied in 294 experiments on 42 neutrophils. As shown in Figure 12, the rheological properties of neutrophils are strongly dependent on the osmolality of the suspending medium. The values of K_1, K_2, and μ increased nearly exponentially with increasing osmolality of the suspension medium from 200 mOsm to 660 mOsm. When the osmolality was reduced from 200 mOsm to 150 mOsm, K_2 increased slightly, but K_1 and μ did not show significant changes. In suspending media with an osmolality of 100 mOsm or less, the deformation of the neutrophils in response to aspiration pressure was reduced as the cell became swollen to approach a

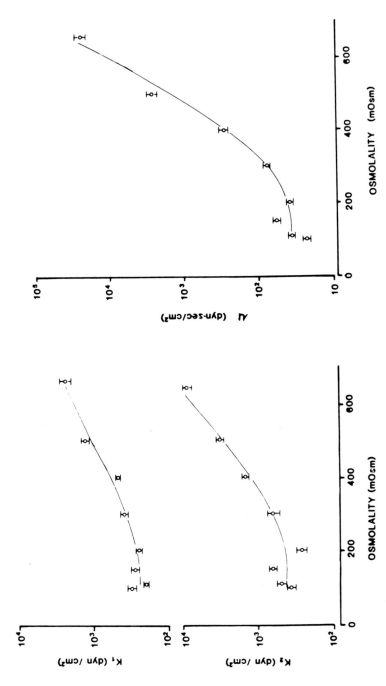

Fig. 12. The viscoelastic coefficients K_1, K_2, and μ for human neutrophils as a function of osmolality of the suspension media at room temperature and pH = 7.4. Vertical bars represent standard errors of the mean.

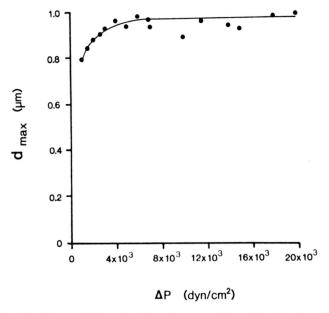

Fig. 13. Steady-state deformation d_{max} of a neutrophil as a function of pressure amplitude ΔP. Note that for aspiration pressures greater than 4,000 dyn/cm^2 the cell did not show any significant increase in d_{max}.

smooth sphere. At these low osmolalities, however, the data were variable among different cells because some of the neutrophils underwent lysis following hypotonic swelling while others were fully or only partially swollen. The speed of Brownian motion of granules in the neutrophils varied inversely with osmolality.

When neutrophils were tested in a solution at 80 mOsm, the deformation of the unlysed cells rapidly reached a steady state (d_{max}) with no significant creep. As shown in Figure 13, d_{max} of a neutrophil swollen in such an hyposmotic medium cannot be increased much further even with applied stresses up to 20,000 dynes/cm^2. Neutrophils that were tested in a 50 mOsm medium did not show any measurable deformation for all aspiration pressures that were applied without rupturing the membrane.

Colchicine. Neutrophils in EDTA were incubated with colchicine (30 μg/ml) for 30 min. The mechanical properties of such treated WBCs were then compared with untreated controls run at the same time. Colchicine treatment did not alter the value of K_1, but it resulted in significant decreases of both K_2 (to 55% of control) and μ (to 10% of control).

Rheological Properties of Neutrophils in the Active State

These studies were performed on neutrophils obtained from fresh heparinized blood samples and suspended in a Ringer's-albumin solution (Ca^{2+} concentration = 2 mM) without EDTA. These WBCs deform spontaneously without an external force acting on them other than a uniform hydrostatic pressure. They develop membrane projections (protopods) which lead to irregular cell shapes, as described by Lichtman et al [26].

The rheological properties of the main cell body and the protopod were compared by measuring their deformation responses to the same micropipette aspiration pressure. Following the application of an aspiration pressure ΔP of 800 dynes/cm^2, the resulting deformation $d(t)$ at the main cell body (Fig. 14, open symbols) showed the typical nonlinear creep, and the deformations at different points on the main cell body were very similar, as in the passive state. In response to the same ΔP of 800 dynes/cm^2, the deformation in the region of the protopod was significantly smaller (Fig. 14, filled symbols) and showed very little creep. When an aspirated protopod rejoined the main body of the cell (Fig. 15), a prominent nonlinear creep developed concurrently as a steady aspiration pressure was maintained ($t > 6$ sec in Fig. 15).

During the first hour following the suspension of WBCs from heparinized blood in Ringer's-albumin in the absence of EDTA, there were neutrophils that retained their spherical shape without protopod formation. Micropipette tests on these neutrophils in Ringer's-albumin without protopods indicate that their viscoelastic coefficients were already significantly higher than the corresponding values obtained from neutrophils in EDTA (Table V).

Filterability of Leukocytes Through Polycarbonate Sieves

When cell-free Ringer's-albumin solution was pumped through polycarbonate sieve at a constant flow rate, the filtration pressure P_0 rose to attain a constant level within approximately 1 sec (Fig. 16). When a suspension of RBCs at a hematocrit of 5%, with very few WBCs, was pumped through the same sieve at the same flow rate, the filtration pressure rose first to attain an initial, quasisteady pressure reading P_i in less than 2 sec, and then it showed a gradual upward slope with continued filtration at a constant flow rate. The value of P_i varied linearly with hematocrit up to 25%. At a given hematocrit, eg, 5%, P_i was not significantly affected by WBC concentration [WBC] up to 12,000/mm^3. An increase in [WBC], however, caused a significant increase in the rate of pressure rise following the attainment of P_i. The pressure rise following

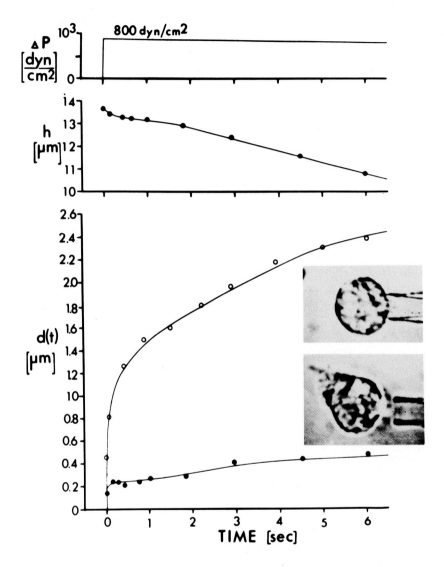

Fig. 14. Micropipette aspiration pressure ΔP and deformation $d(t)$ for a neutrophil which was aspirated at the main cell body (open circles) and at a small protopod (dots). h is the length of the major axis of the cell (from [16]).

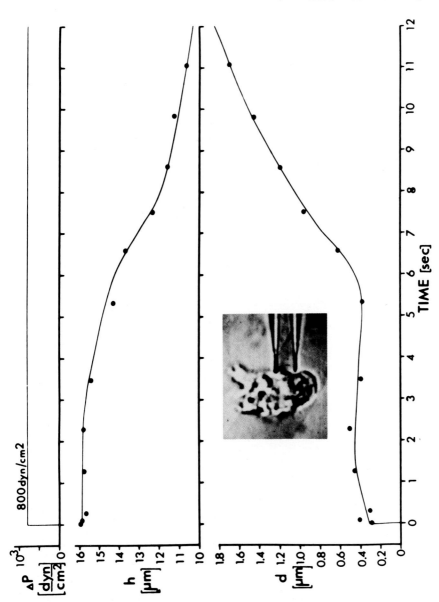

Fig. 15. The aspiration pressure ΔP, total length of main cell plus protopod h (measured vertically in the inset photograph), and deformation d into the pipette. At about t = 4 sec the protopod spontaneously began to retreat (from [16]).

TABLE V. Comparison of Viscoelastic Properties of Human Neutrophils in Heparin-Ringer's and EDTA-Saline[a]

	Heparin-Ringer's (n = 12)	ESTA-saline (n = 9)
K_1 (dynes/cm^2)	258 ± 21	166 ± 32
K_2 (dynes/cm^2)	352 ± 30	269 ± 33
μ (dynes-sec/cm^2)	119 ± 33	66 ± 10

[a]All neutrophils studied had an overall spherical shape. Temperature = 35–37°C. Values are means ± SEM. All three viscoelastic coefficients are significantly higher in the heparin-Ringer group ($P < .01$).

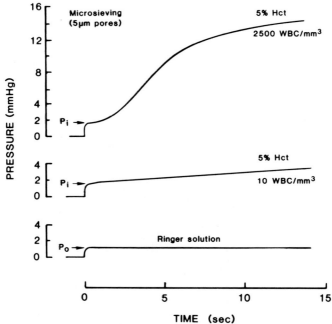

Fig. 16. Pressure-time curves (from bottom to top) of a cell-free Ringer's solution, a cell suspension containing 5% hematocrit and 10 WBCs/mm^3, and a cell suspension containing 5% hematocrit and 2,500 WBCs/mm^3, filtered through 5-μm polycarbonate sieves at a constant flow rate.

P_i generally shows first a fast phase which is followed by a slower phase. The slopes of the fast and slow phases of pressure rise were divided by P_0 and designated k_1 and k_2 (in sec^{-1}), respectively. Both k_1 and k_2 are strongly dependent on [WBC]. When [WBC] was kept in a low range of 250–600/mm^3 and the hematocrit was elevated, k_1 was also found to vary directly with the RBC concentration, [RBC]. The relative influences of

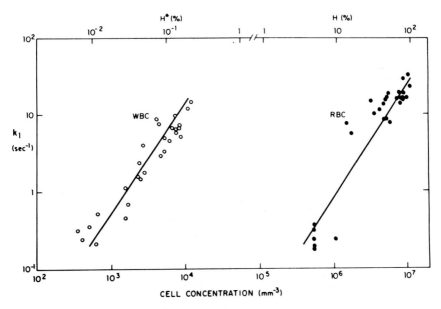

Fig. 17. Double logarithmic plot comparing the effects of leukocyte concentration [WBC] and erythrocyte concentration [RBC] on the slope of the fast phase of pressure rise (k_1). The top abscissa scale is drawn to show the leukocrit (H*) and hematocrit (H) in volume percent. In order to achieve a given value of k_1, the [RBC] required is approximately 700 times higher than [WBC] (from [17]).

[RBC] and [WBC] on k_1 are compared in the log-log plot of Figure 17. In order to obtain the same k_1, the [RBC] needed is approximately 700–fold higher than [WBC]. That is, each WBC is equivalent to approximately 700 RBCs in causing the fast phase of the pressure rise.

The obstruction of the filter pores by WBCs can be shown by scanning and transmission electron microscopy. Transmission electron microscopy of thin sections cut parallel to the filter pore allows the direct demonstration of pore plugging by WBCs (Fig. 18). In Figure 19, the obstruction of the exit side of filter pores by an adhering WBC and several RBCs is visible under scanning electron microscopy.

DISCUSSION

When human neutrophils in EDTA are subjected to small deformations by micropipette aspiration, their viscoelastic behavior can be modeled with a three-element standard solid material. The agreement between the model and the data is good for small strains with $d(t) < 0.9 \mu m$. For larger

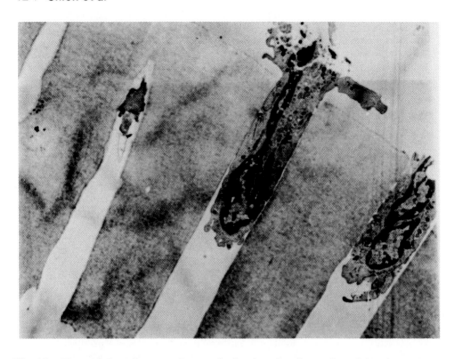

Fig. 18. Transmission electron micrograph showing the obstruction of filter pores by leukocytes. The upstream side of the filter is at the upper left. ×3,200 (from [17]).

deformations, when some of the assumptions made in the modeling [13] are not applicable, the surface displacements computed from the model overestimate the measured values. The empirical coefficients K_1, K_2, and μ can be used to explain the recorded deformation history d(t). At very short times after the initiation of deformation, the neutrophils behave like elastic bodies with a shear modulus equal to $(K_1 + K_2)/2$. The slope of d(t) at t = 0(+) is proportional to $K_2^2/[(K_1 + K_2)^2\mu]$, and the maximum displacement for long times is proportional to $1/K_1$.

The standard solid model gives mean properties of the cell, including the stress contributions in all cytoplasmic structures and the membrane. In an isotonic medium at physiological pH, when there exists a large excess membrane area (Table I), the contribution of the cell geometry to the viscoelastic behavior of the neutrophil during small deformations is probably negligible [13]. The response is dominated by the cell interior which is composed of materials with high values of viscoelastic coefficients as compared to those for the cell membrane undergoing deformation at constant area. This is to be contrasted with the response of normal RBCs to small deformation by micropipette aspiration. The normal RBC interior

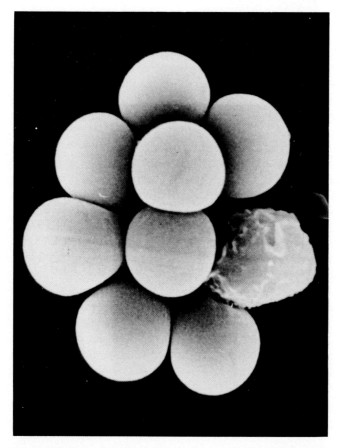

Fig. 19. Scanning electron micrograph showing the obstruction of the exit side of a filter pore by adherent cells (one leukocyte and eight erythrocytes) forming a bouquet. ×6,600 (from [17]).

has no elastic modulus and a low intracellular viscosity [24, 27], and the response to small deformations reflects primarily the rheological behavior of the membrane [21, 28, 29]. It is only when the intracellular viscoelastic coefficients become abnormally elevated, eg, in sickle cells following deoxygenation [27], that the deformational behavior of RBCs would become dominated by the intracellular content.

Changes in passive rheological properties of neutrophils following alterations in physicochemical conditions can be demonstrated by the micropipette method. The effect of colchicine on the passive viscoelastic coefficients of neutrophils is of special interest because of its action on the microtubules, which are part of the cytoskeletal apparatus playing an important role in providing the mechanical support [30, 31]. The selective decreases in K_2 and μ without a significant alteration in K_1 following colchicine treatment suggest that the integrity of the microtubules serves to provide the viscoelastic resistance of neutrophils to externally applied deforming stress. Furthermore, these results suggest that the Maxwell element (K_2 in series with μ) of our rheological model has a biochemical counterpart represented by the microtubules. The roles of the microfilaments in the cytoskeletal apparatus in affecting the WBC deformational behaviors are currently under investigation.

Variations in temperature from 9°C to 40°C do not have significant influences on K_1 and K_2. The value of μ varies inversely with temperature; this is similar to the temperature dependence of viscosity seen in other liquids and RBCs [32, 33]. Following an increase in pH to 8.4, the neutrophils show increases in K_1 and μ, and these rheological changes are associated with swelling of the cell and its granules and nucleus. The loss of cell membrane foldings with neutrophil swelling would reduce the excessive membrane surface area available for deformation and may contribute to the observed changes in rheological properties. There is also the possibility that high pH rigidifies some components of the neutrophil interior, eg, the cytoskeletal apparatus. It is interesting to note that reduction of pH to a value as low as 4.9 does not affect the rheological properties of neutrophils. Micropipette tests on RBCs have shown progressive increases in membrane shear modulus and viscosity with pH ≤ 5.9 [34, 35]. These results indicate that the human RBC membrane is more sensitive to acidosis than the neutrophils.

An elevation in the osmolality of the suspending media causes marked increases of all three coefficients of neutrophils, probably reflecting the increase in solid concentration due to hyperosmotic shrinkage and cellular dehydration. In media with normal osmolality (~ 300 mOsm), the μ value of neutrophils is about 10^2 poises; in an hyperosmotic medium of 660 mOsm, the μ value of neutrophils increases by approximately 300-fold to 3×10^4 poises. The elastic moduli K_1 and K_2 increase by lesser degrees. When the osmolality is reduced from 300 to 200 mOsm, there is a slight decrease in μ, as the neutrophil interior becomes diluted. RBCs begin to exhibit hemolysis when the osmolality of the medium is reduced below 150 mOsm [36, 37]. In comparison, the neutrophils have more tolerance to hyposmotic treatment, as most neutrophils are stable in 100 mOsm medium without rupturing. The cytoplasm of such neutrophils is diluted with water, and the cell has an almost spherical shape with unfolded membrane. The overall rheological properties of the neutrophil in response to variations in osmolality reflect a balance between the influences of the intracellular solid concentration and the geometric relation between membrane surface area and cell volume. At low osmolalities near 200 mOsm, the relative constancy of the viscoelastic coefficients probably results from a balance of the decrease in solid concentration and the increase in sphericity, resulting from the reduction in osmolality. At osmolalities of 100 mOsm or lower, when all the surface foldings in most cells disappear and the neutrophil becomes a smooth sphere, the cell is essentially nondeformable despite the increase in fluidity of the cell interior.

Previous workers [2, 3] have applied micropipette suction to study human leukocytes, but the design of their experiments is different from ours. They used pipettes with larger inner diameter ($\simeq 5$ μm) to aspirate the entire cell inside the pipette and employed the pressure needed for this process as a measure of the elasticity of the cell. Deformation of the entire WBC may involve the nucleus and other structures with higher viscoelastic coefficients, as well as the unfolding of the cell membrane, and there may be nonlinearity of the coefficients in various components of the WBC. With the use of a three-element model in an one-dimensional strain analysis, Bagge et al [38] deduced from such large deformation studies that $K_1 \simeq 3$ dynes/cm, $K_2 \simeq 30$ dynes/cm and $\mu \simeq 60$ dynes-sec/cm. These results cannot be compared directly with those obtained in the present study, but they probably represent higher viscoelastic coefficients. The time constant of the WBC deformation process, which can be expressed as $\mu(K_1 + K_2)/K_1K_2$, is on the order of 0.6 sec for the small deformation of WBC in the present study. In contrast, the time constant in large deformation of the entire WBC [3] is on the order of 20 sec. This discrepancy illustrates the difference in WBC properties when tested with a small deformation (the present study) and when examined with a large deformation of the whole cell [38].

Our data show that during active deformation the leukocytes' mechanical properties change dramatically in the region of the protopod. Biochemical analysis of extracts of horse leukocytes [39], guinea pig granulocytes [40], and rabbit macrophages [41] indicates that actinlike and myosinlike proteins are present, as well as other proteins which interact to control their polymerization. Gelation of these proteins appears to generate the formation of the protopod and the change in the mechanical properties of the cell. The granules, mitochondria, and nucleus, which together account for 37% of the cell volume, are excluded from the protopod. The formation of the protopod is initiated at a region near the membrane and progresses centrally until the process is reversed by mechanisms as yet not completely understood. Geometrically, the protopod is an arrangement of sheetlike projections, which have an unfolded membrane. A theoretical analysis describing the mechanical phenomena of active deformation is presented in an accompanying paper [43].

When cell suspensions are pumped through polycarbonate sieves with 5-μm pore diameters at a constant flow rate, the filtration pressure shows an immediate rise within 1–2 sec to a level (P_i) which is primarily determined by the RBC concentration [23]; the progressive rise in P following the attainment of P_i is strongly dependent on WBC concentration. This secondary rise in P shows an initial fast phase and a later slow phase, with a transition period between these two phases. The initial, fast phase of pressure rise following P_i probably reflects the plugging of filter pores by some fractions of the entering WBCs and RBCs, as well as the dislodgement of some of the plugged cells by the rising filtration pressure [44]. The later, slow rise in P reflects that small fractions of the entering cells remain plugged in the filter essentially permanently. The plugging of filter pores by WBCs is demonstrated by SEM and TEM (Figs. 18, 19).

An elevation of either [WBC] or [RBC] leads to increases in the slopes for both the initial and late phases of pressure rise. The effects of WBCs, however, are markedly stronger. The [RBC] must be nearly three orders of magnitude higher than the [WBC] in order to generate the same slope for the fast phase of pressure rise (k_1). That is, one WBC is approximately equivalent to 700 RBCs in causing the plugging of 5-μm pores. This is in keeping with the finding that the cellular viscosity of WBCs is three orders of magnitude higher than the RBC content (Table IV). Another interesting point to be noted is that the normal [RBC] of $5 \times 10^6/mm^3$ is approximately 700 times the normal [WBC] of $7 \times 10^3/mm^3$. Even completely packed red cells with a hematocrit of 100%, where the [RBC]= $1.1 \times 10^7/mm^3$, are only equivalent to approximately 10^4 WBC/mm^3 in causing the plugging of 5-μm channels. These results illustrate the potential importance of increased [WBC] in causing microvascular obstruction.

The findings of these morphometric and rheological investigations have considerable implications concerning the functional roles of WBCs in health and disease states. The WBCs are comparable to RBCs in their diameters, but WBCs have larger volumes due to their spherical shape. Although the WBCs have an overall spherical contour, the presence of membrane foldings provides them with excess membrane areas greater than that in the discoid erythrocytes. From a geometric point of view, as far as the cell volume and membrane area are concerned, the WBCs can be deformed to pass through as narrow a channel (2.7 μm or less) as the RBCs (Table I). The relative inability of WBCs to pass through 5-μm channels, as compared to RBCs, must be attributed to the difference in their viscoelastic properties. The passive elastic properties, as reflected by the deforming stress required to attain a given degree of steady-state small deformation (Fig. 9), differ only by a factor of 4, but the WBCs have a cellular viscosity more than 1,000-fold higher than the intracellular viscosity of normal RBC (Tables III, IV). The difference in cellular viscosity and the time constant of deformation may be an important factor

EFFECT OF WBC ON RBC DISTRIBUTION

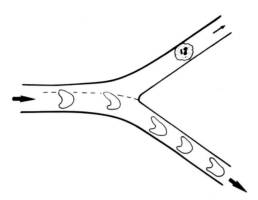

Fig. 20.   Schematic drawing of a divergent capillary bifurcation, showing the influence of the entry of a WBC into one branch on the flow rates (indicated by the relative sizes of the arrows), the partitioning of the streamlines in the parent vessel (indicated by the dashed line), and the preferential entry of the subsequently arriving RBCs into the other branch.

in causing the relative difficulty of WBC passage through 5-μm channels. Another possible contributing factor is that the unfolding of the excess membrane area in WBCs may offer additional resistance to WBC deformation. Although the geometric features of the whole WBC indicate that they can be deformed to a cylindrical sausage shape with a diameter smaller than 2.7 μm, such deformation cannot be attained because of the presence of the nucleus which is much less deformable than the cell cytoplasm [unpublished observation]. In particular the ability of lymphocytes to deform is limited because of the relatively large cell nucleus (Table II) and the small excess nuclear membrane area (Table I).

The low deformability and the slow transit of WBCs in narrow vessels influence the flow dynamics in a microcirculatory network. In a divergent bifurcation, the entry of a WBC into one branch will raise the flow resistance and reduce the flow rate in that branch (Fig. 20). As a result, the RBCs arriving subsequently at the branch point will be diverted into the other branch until the resistances and flow rates become balanced again [45]. In a convergent branch, eg, postcapillary venules, the interaction between a WBC and a RBC arriving from the two feeding capillaries may cause the WBC to be in close contact with the venular endothelium (Fig. 21) and rolling along the wall [7, 46]. Therefore, the WBCs tend to occupy a position where they can most readily perceive chemotactic signal from the extravascular space and emigrate across the vessel wall when

EFFECT OF RBC ON WBC–ENDOTHELIUM INTERACTION

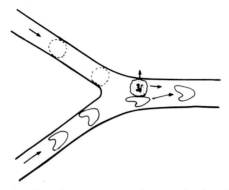

Fig. 21. Schematic drawing of a convergent microvascular branch point, showing the influence of RBCs on the WBC, causing the WBC to remain in a flow stream very close to the endothelium.

needed. The margination of WBCs will be further enhanced when there is RBC aggregation [47, 48].

The strong influence of [WBC] on filter plugging indicates that an abnormal elevation of [WBC], eg, in leukemia, may lead to microvascular obstruction [49]. In addition to total [WBC], other important factors to be considered are the relative proportion of the various types of leukocytes, their rheological properties, and their adhesiveness to the walls of the flowing channel. The deformability and adhesiveness of the leukocytes may be adversely affected in hemorrhage and other low flow states, thus enhancing the probability of microvascular obstruction by leukocytes [5, 50].

SUMMARY

The viscoelastic properties of leukocytes (WBCs) during small deformation were determined by micropipette aspiration. The passive deformation behavior of neutrophils suspended in a Ca^{2+}-free medium in response to a step aspiration pressure consists of an initial rapid, elastic response followed by a creep displacement. These time-dependent responses can be modeled by a viscoelastic solid in which an elastic element (K_1) is in parallel with a Maxwell element composed of another elastic element (K_2) in series with a viscous element (μ). Variations in temperature (9–40°C) cause an inverse change in μ, but have no effects on K_1 and K_2. All three coefficients are not affected by decreases in pH down to 5.4; with pH $\geq$

7.8, however, K_1 and μ increase. Increases in osmolality cause a rise in all three coefficients, especially μ. Colchicine treatment results in selective decreases in μ and K_2 without affecting K_1. B lymphocytes have viscoelastic coefficients similar to those of neutrophils, but T lymphocytes have higher values for these coefficients. In the presence of 2 mM Ca^{2+}, the neutrophils have higher viscoelastic coefficients than in Ca^{2+}-free medium, and they form protopods which have greater resistance than the main cell body to deformation by micropipette aspiration. Morphometric analysis shows that WBCs have large excess membrane area due to the presence of membrane foldings, which facilitate WBC deformation at constant area. During filtration through 5-μm sieves, WBCs are much more prone to pore plugging than erythrocytes because of their higher cellular viscosity and the presence of nucleus. The rheological properties of WBCs have significant implications in their functions and flow dynamics in the microcirculation.

ACKNOWLEDGMENTS

This work was supported by U.S.P.H.S. Research Grant HL-16851 and NRSA HL-07114 from the National Heart, Lung and Blood Institute.

REFERENCES

1. Chien S (1982): Rheology in the microcirculation in normal and low flow states. Adv Shock Res 8:71–90.
2. Lichtman MA (1973): Rheology of leukocytes, leukocyte suspensions and blood in leukemia. J Clin Invest 52:350–358.
3. Miller ME, Myers KA (1975): Cellular deformability of the human peripheral blood polymorphonuclear leukocyte: Method of study, normal variation, and effects of physical and chemical alterations. J Reticuloendothel Soc 18:337–345.
4. Adell R, Skalak R, Brånemark PI (1970): A preliminary study of rheology of granulocytes. Blut 21:91–105.
5. Bagge U, Amundson B, Lauritzen C (1980): White blood cell deformability and plugging of skeletal muscle capillaries in hemorrhagic shock. Acta Physiol Scand 108:159–163.
6. Bagge U, Brånemark PI (1977): White blood cell rheology. An intravital study in man. Adv Microcirc 7:1–17.
7. Schmid-Schönbein GW, Usami S, Skalak R, Chien S (1980): The interaction of leukocytes and erythrocytes in capillary and postcapillary vessels. Microvasc Res 19:45–70.
8. Marchesi VT, Florey HW (1960): Electron micrographic observations on the emigration of leukocytes. Quart J Exp Physiol 45:343–347.
9. Gallin JI, Quie PC (eds) (1978): Leukocyte Chemotaxis: Methods, Physiology and Clinical Implications. New York: Raven Press.

10. Wilkinson PC (1974): Chemotaxis and Inflammation. Edinburgh and London: Churchill Livingstone.
11. Zigmoid SH (1978): Chemotaxis by polymorphonuclear leukocytes. J Cell Biol 77:269–287.
12. Elsbach P (1974): Phagocytosis. In Zweifach BW, Grant L, McCluskey RT (eds): "The Inflammatory Process." New York: Academic Press, Inc., 2nd ed., ch 1, pp 363–410.
13. Schmid-Schönbein GW, Sung KLP, Tözeren H, Skalak R, Chien S (1981): Passive mechanical properties of human leukocytes. Biophys J 36:243–256.
14. Schmid-Schönbein GW, Shih YY, Chien S (1980): Morphometry of human leukocytes. Blood 56:866–875.
15. Sung P, Schmid-Schönbein GW, Skalak R, Schuessler GB, Usami S, Chien S (1982): Influence of physicochemical factors on rheology of human neutrophil. Biophys J 39:101–106.
16. Schmid-Schönbein GW, Skalak R, Sung KLP, Chien S (1982): Human leukocytes in the active state. In Bagge U, Born GV, Gaehtgens P (eds): "White Blood Cells, Morphology and Rheology as Related to Function." The Hague: Martinus Nijhoff, pp 21–31.
17. Chien S, Schmalzer EA, Lee MML, Impulluso T, Skalak R (1983): Role of white blood cells in filtration of blood cell suspensions. Biorheology 20:11–27.
18. Simionescu N, Simionescu M (1976): Galloyl glucose of low molecular weight as mordant in electron microscopy. I. Procedure and evidence for mordanting effect. J Cell Biol 70:608–621.
19. Weibel ER (1969): Stereological principles for morphometry in electron microscopic cytology. In Bohrne GH, Danielli JF (eds): "International Review in Cytology." New York and London: Academic Press, Vol 26, pp 235–302.
20. Underwood E (1970): Quantitative Stereology. Reading, Massachusetts: Addison-Wesley Co.
21. Chien S, Sung KLP, Skalak R, Usami S, Tözeren A (1978): Theoretical and experimental studies on viscoelastic properties of erythrocyte membrane. Biophys J 24:463–487.
22. Usami S, Chien S, Bertles JF (1975): Deformability of sickle cells as studied by microsieving. J Lab Clin Med 86:274–279.
23. Schmalzer EA, Skalak R, Usami S, Vayo M, Chien S (1983): Influence of red cell concentration on filtration of blood cell suspensions. Biorheology 20:29–40.
24. Cokelet GR, Meiselman HJ (1968): Rheological comparison of hemoglobin solutions and erythrocyte suspensions. Science 162:275–277.
25. Chien S, Usami S, Bertles JF (1970): Abnormal rheology of oxygenated blood in sickle cell anemia. J Clin Invest 49:623–634.
26. Lichtman MA, Santillo PA, Kearney EA, Roberts GW, Weed RI (1976): The shape and surface morphology of human leukocytes in vitro: Effect of temperature, metabolic inhibitors and agents that influence membrane structure. Blood Cells 2:507–531.
27. Chien S, King RG, Kaperonis AA, Usami S (1982): Viscoelastic properties of sickle cells and hemoglobin. Blood Cells 8:53–64.
28. Hochmuth RM, Berk DA, Wiles HC (1983): Viscous flow of cytoplasm and red cell membrane: Membrane recovery and tether contraction. Ann NY Acad Sci (in press).
29. Tözeren H, Chien S, Tözeren A (1983): Estimation of viscous dissipation inside an erythrocyte during aspirational entry into a micropipette. Biophys J (submitted).
30. Howard TH, Casella J, Lin S (1981): Correlation of the biologic effects and binding of cytochalasins to human polymorphonuclear leukocytes. Blood 57:399–405.

31. Cain H, Kraus B (1981): Cytoskeleton in cells of the mononuclear phagocyte system. Immunofluorescence microscopic and electron microscopic studies. Virchows Arch Cell Pathol 36:159–176.
32. Chemical Rubber Publishing Co. (1962): Handbook of Chemistry and Physics. Cleveland, Ohio: Chemical Rubber Publishing Co., p 2257.
33. Hochmuth RM, Buxbaum KL, Evans EA (1980): Temperature dependence of the viscoelastic recovery of red cell membrane. Biophys J 29:177–182.
34. Crandall ED, Critz AM, Osher AS, Keljo DJ, Forster RE (1978): Influence of pH on elastic deformability of the human erythrocyte membrane. Am J Physiol 235:C269–278.
35. Sung KLP, Chien S (1983): Influence of physicochemical factors on viscoelasticity of red cell membrane. Fed Proc 42:605.
36. Dodge JT, Mitchell C, Hanahan DJ (1963): The preparation and chemical characterization of hemoglobin-free ghosts of human erythrocytes. Arch Biochem 110:119–130.
37. Katchalsky A, Keden O, Klibansky C, DeVries A (1960): Rheological considerations of the haemolysing red blood cell. In Copley AL, Stainsby G (eds): "Flow Properties of Blood and Other Biological Systems." New York: Pergamon Press, pp 155–164.
38. Bagge U, Skalak R, Attefors R (1977): Granulocyte rheology. Adv Microcirc 7:29–48.
39. Shibata N, Takubo T, Senda N (1979): Ca^{2+}-sensitive contractile protein from leucocytes. In Hatano S, Ishikawa H, Sato H (eds): "Cell Motility: Molecules and Organelle." Baltimore: University Park Press, pp 13–31.
40. Stossel TP, Pollard TD (1973): Myosin in polymorphonuclear leukocytes. J Biol Chem 248:8288–8294.
41. Hartwig JH, Stossel TP (1975): Isolation and properties of actin, myosin, and a new actin binding protein in rabbit alveolar macrophages. J Biol Chem 250:5696–5705.
42. Brotschi EA, Hartwig JH, Stossel TP (1978): The gelation of actin by actin-binding protein. J Biol Chem 253:8988–8993.
43. Skalak R, Chien S, Schmid-Schönbein GW (1983): Viscoelastic deformation of WBC: Theory and analysis. This Symposium.
44. Skalak R, Impelluso T, Schmalzer EA, Chien S (1983): Theoretical modeling of filtration of blood cell suspensions. Biorheology 20:41–56.
45. Schmid-Schönbein GW, Usami S, Skalak R, Chien S (1980): Cell distribution in capillary networks. Microvasc Res 19:18–44.
46. Bagge U, Karlsson R (1980): Maintenance of white blood cell margination at the passage through small venular junctions. Microvasc Res 20:92–95.
47. Nobis U, Gaehtgens P (1981): Rheology of white blood cells during blood flow through narrow tubes. Bibl Anat 20:211–214.
48. Chien S, Usami S, Skalak R (1983): Blood flow in small tubes. In Renkin EM, Michel C (eds): "Handbook of Physiology: Microcirculation." Washington, DC: Am Physiol Soc (in press)
49. Preston FE, Sokol RJ, Lileyman JS, Winfield DA, Blackburn EK (1978): Cellular hyperviscosity as a cause of neurological symptoms in leukemia. Br Med J 1:476–8.
50. Braide M, Amundsson B, Chien S, Bagge U (1983): Quantitative studies on the influence of leukocytes on the vascular resistance in a skeletal muscle preparation. Microvasc Res (submitted).

White Cell Mechanics: Basic Science and
Clinical Aspects, pages 53–71
© 1984 Alan R. Liss, Inc., 150 Fifth Avenue, New York, NY 10011

Structural Model for Passive Granulocyte Behaviour Based on Mechanical Deformation and Recovery After Deformation Tests

Evan A. Evans

Department of Pathology, University of British Columbia, Vancouver, BC Canada V6T 1W5

INTRODUCTION

The ability to repeatedly deform, enter, and transit small capillaries in the microcirculation without causing occlusion of the vessel, damage to the cell, or damage to the vessel wall is the minimum but essential functional requirement for any blood cell. In addition, white cells possess the capability of penetration through the vascular endothelium and movement into the extravascular space. Unlike its companion cell, the erythrocyte, the white cell can be self-motive and thus is not merely a passive material structure. In order to better understand the in vivo rheological behaviour of white cells, it is important to determine the deformation and activity of white cells in response to the physical forces that these cells experience in the circulation. Although rheological properties of the companion erythrocytes have been well investigated, the rheological properties of white cells are not as well established and appear to be very sensitive to the environment. Here, we review our studies [Evans and Kukan, 1983] of the time-dependent deformability, recovery, and activity of granulocytic, white blood cells. Large deformation and recovery characteristics after deformation of neutrophils have been evaluated with micropipette aspiration techniques. Based on our results, we suggest a simple structural model to represent the rheological behaviour of the granulocyte, which can be used to quantitate the physiochemical state of cellular components. In addition, we describe micromanipulation methods to quantitate the phagocytic and surface recognition properties of white cells; we have used these methods to evaluate the potential activation of white cells by mechanical stimuli.

Recognizing the functional requirements for white cells in the circulation, many investigators have studied the behaviour of granulocytes in mechanical tests, although white cell mechanical properties have only been cursorily investigated in comparison to red cells. The investigations have ranged from evaluation of the effects of bulk fluid shear stress on white cell function and survival in viscometers [Martin et al, 1979], to single cell studies of white cell passage through microcirculatory beds [Bagge, 1976], through micropore filters [Lichtman and Kearney, 1976], and into micropipettes [Lichtman and Weed, 1972; Miller and Myers, 1975; Bagge, 1976; Schmid-Schönbein et al, 1981]. These studies on white cells have focused on two extremes of deformability: (1) the instantaneous response of the cell to an applied pressure (less than 1 to 2 seconds [Schmid-Schönbein et al, 1981]); and (2) the pressure drop necessary for white cells to completely transit a filter pore or enter a micropipette [Lichtman and Kearney, 1976; Lichtman and Weed, 1972; Miller and Myers, 1975; Bagge, 1976]. However, the detailed dynamics of large deformation and flow into a small-calibre channel, plus recovery after deformation, for the white cell have been relatively neglected. The significance of the difference between large and small deformation properties of the white cells is apparent from the comparison of the respective correlations of rheological models with the observed behaviour. For example, correlations of a "standard viscoelastic solid" model with the data for the transit of white cells through a tapered channel [Bagge, 1976] give entirely different results for the model parameters than those obtained for the correlation of this same viscoelastic model with the small deformation behaviour [Schmid-Schönbein et al, 1981]. Specifically, the time constant for viscoelastic response and recovery after deformation is estimated to be on the order of 50–80 seconds in the tapered-channel experiments, whereas it is only 1–2 seconds for the small-deformation experiments. Also, the "standard viscoelastic solid" model predicts that a static (time-independent) level of deformation should be effectively reached for times longer than 2–3 time constants. This is not consistent with our experimental results [Evans and Kukan, 1983] which show that the deformation process is a continual flow in response to a constant stress over time periods of several minutes and can eventually lead to total aspiration of the white cell in our pipet suction tests. Furthermore, there appears to be an enormous variability in the magnitudes of suction pressures required to aspirate white cells into micropipets or to transit small filter pores; for instance, with channel diameters of about 3.5 μm, the pressures required to totally aspirate normal granulocytes range from 1,000 to 10,000 dynes/cm^2 in the previously given references, whereas in our experiments it takes only about 200–300 dynes/cm^2. Hence, the rheological, constitutive behaviour of the white cell is not well established.

It has been generally recognized and experimentally verified that the capability of the white cell to deform is enhanced by the large reservoir of membrane material which is present as surface folds, and that this reservoir represents approximately 80% excess area over a sphere of equivalent volume [Bagge, 1976; Schmid-Schönbein et al, 1980]. These data imply a minimum cylindrical diameter of 2.6 μm for complete passage of a granulocyte through a capillary tube. Also, it has been observed that the white cell is capable of rapid recovery to a nearly perfect, spherical shape after large deformations [Bagge, 1976] provided that the cell does not become active. Furthermore, it was noted that cell activity alters the deformability properties [Schmid-Schönbein et al, 1981] and that removal of calcium by the addition of ethylenediamine tetra-acetic acid (EDTA) maintains the white cell in a passive state.

METHODS OF PROCEDURE

Micromechanical experiments on individual white cells were performed with micropipette aspiration methods similar to those used by Schmid-Schönbein et al [1981], except that the white cell aspiration in our experiments was allowed to proceed for long time periods (5 minutes) as opposed to 1–2 seconds in the previous tests. The procedure involved two micropipettes: one pipette was used to convey a white cell into position for rapid aspiration by a second pipette which had a fixed suction pressure in the range of 200–700 dynes/cm^2. The exposure to the suction pressure in the second, measurement pipette occurred within less than 0.05 seconds. Subsequent to the 5-minute exposure of the white cell to the aspiration pressure, the pressure was set to zero, and the recovery after deformation of the white cell was observed. The progress of the experiment was recorded on videotape, examples of which are shown in Figure 1. Analysis of the data involved measurement of the length inside the pipette throughout the aspiration phase and the major dimension of the cell during the recovery phase, schematically illustrated in Figure 2. An example of this data is shown in Figure 3. In general, the cell response in the pipette experiment was characterised by three time domains: the first was the passive response to the aspiration pressure; this was followed by an obvious transition from the passive to active, motile cellular state where

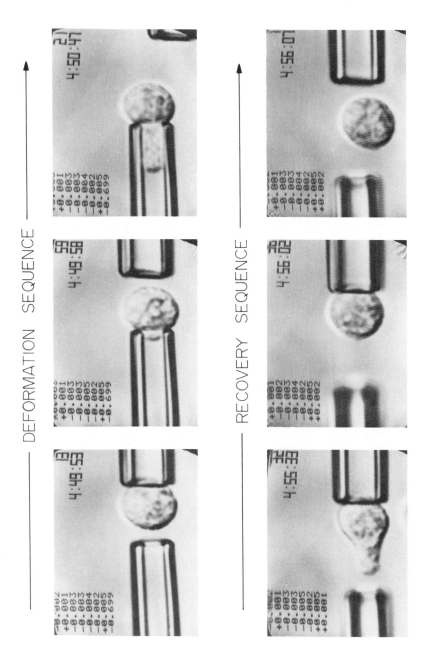

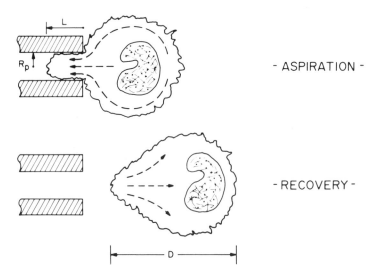

Fig. 2. Schematic illustration of the flow into the pipette during the aspiration phase and the recovery flow following expulsion from the pipette. Shown here are the length L and D which characterize the aspiration and recovery phases, respectively (taken from Evans and Kukan [1973]).

the white cell exhibited erratic length changes in the micropipette as shown in Figure 3; and, finally, a rapid recovery after the suction pressure had been zeroed. Some of the recovery after deformation occurred inside the pipette followed by complete recovery to the spherical state outside the pipette as shown in Figure 1. We found that the white cells were most deformable in blood type-compatible plasma and exhibited no adhesion to the walls of the micropipette. Three specific micropipette sizes were used: 2.7 μm, 3.5 μm, and 5.2 μm. The internal diameters of these pipettes were determined from the insertion depth of a microneedle which had been calibrated with the use of a scanning electron microscope. Pressures were recorded through a continuous water manometer system to a sensitive digital pressure transducer with a resolution of a few microatmospheres of pressure. Neutrophils were selected at random from an extremely low

Fig. 1. Sequence of video micrographs of granulocyte deformation and recovery after deformation. The pipette diameter is 3.5 μm; the suction pressure is 700 dynes/cm^2 (given by the bottom digital voltmeter reading); and the time is in the upper right hand corner of the video image (taken from Evans and Kukan [1983]).

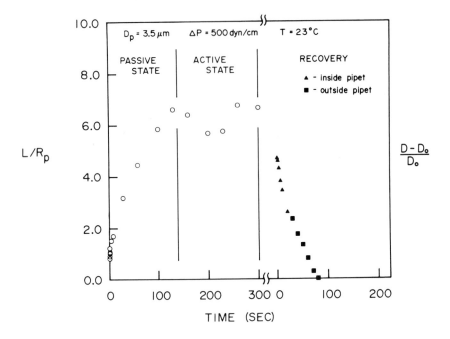

Fig. 3. Data for a single granulocyte aspiration-and-recovery experiment. The aspiration phase, plotted as the open circles, often exhibited a definite transition for the white cell from a passive to active state. The recovery-after-deformation phase, plotted as the closed triangles and closed squares, began initially in the pipette and continued outside the pipette after the cell was extracted by a second pipette (taken from Evans and Kukan [1983]).

hematocrit blood sample in the microchamber. The selection was facilitated with the use of a differential interference contrast optical system; the identification was based on the observation of small-sized, densely packed granules within the cell and (as best as could be detected) a lobulated nucleus. The granulocytes that were selected fell into a reasonably narrow size category of 8.5 to 9.4 μm in diameter.

To test the agglutinating and phagocytic properties of granulocytes, we have developed micromanipulative methods where the exposure of the white cell to stimuli in solution and on particle surfaces can be controlled. These methods were developed in order to directly observe cell deformation and to measure cell activation forces in white cell-surface interactions. The approach utilizes three micropipettes and two adjacent microchambers separated by a small air gap. One of the pipettes is a large transfer pipette (about 20 μm in diameter) which spans the air gap and into which a white cell or test particle is inserted by one of the small micropipettes.

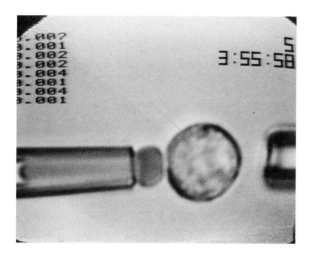

Fig. 4. Video micrograph of granulocyte adhesion to a wheat germ agglutinin (WGA) coated red blood cell. The red cell was first coated with WGA, transferred to the chamber which contained the granulocyte in WGA-free solution, then the red cell and white cell were maneuvered into position and contact allowed to occur. The result was strong adhesion but no further activity.

The microscope stage is translated to leave the transfer pipette with the inserted object in the second microchamber. The object is then withdrawn and maneuvered into proximity of a test cell held by the third pipette. An example of the procedure is as follows: red cells in solution with the plant lectin wheat germ agglutin (WGA) were placed in one chamber; granulocytes were placed in WGA-free solution in the adjacent chamber. A red cell coated with WGA was transferred to the chamber which contained the granulocytes in WGA-free solution. The red cell and a white cell were maneuvered into position and contact occurred. The result is shown in Figure 4: adhesion, no phagocytosis. This procedure was also used to test zymosan granules. The result is shown in Figure 5: adhesion followed by phagocytosis.

The blood cell samples for use in the micromechanical experiments were obtained freshly by finger-prick, immediately diluted into filtered blood type-compatible plasma, and then injected into a small microchamber on the microscope stage. The cell experiments were carried out over a period from 5 to 40 minutes and the results exhibited no dependence on the time period in the chamber. The blood type-compatible plasma was obtained from separated fresh blood samples (20–40 ml) from the same

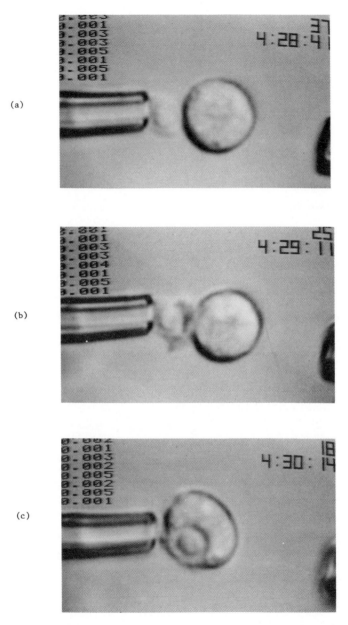

Fig. 5. Sequence (a–c) of video micrographs of granulocyte phagocytosis of a zymosan granule. The procedure was similar to that described in Figure 4; a single zymosan granule was transferred to the chamber which contained the granulocyte, then the bodies were maneuvered into position to permit adhesion and phagocytosis. The rapid time course of the event is demonstrated by the time recording in the upper right hand corner of the video image.

donor with sodium citrate as the anticoagulant. The blood samples were separated by centrifugation; the supernatant plasma was removed and filtered for use in the experiment. The micromechanical tests were carried out at both room temperature (23°C) and at 37°C where the temperatures were maintained by a specially designed heat exchanger for the microscope chamber.

RESULTS

The micromechanical experiments [Evans and Kukan, 1983] involved observation of the large deformation response of the white cell to aspiration by fixed suction pressures of 200, 500, and 700 dynes/cm² with three different-sized pipettes of 2.7, 3.5, and 5.2 μm, respectively. Also, the time for transition from a passive to active, motile state was observed and, finally, the time course of recovery after deformation in the pipette experiments was quantitated. These experiments were carried out at two different temperatures of 23°C and 37°C. Because of the long duration of the aspiration experiment (5 minutes) and large aspiration lengths, the steady flow of a white cell into the micropipette is best plotted as the logarithm of the length to pipet ratio verus the logarithm of time as shown in Figures 6–10. Each curve represents the average of measurements on five to ten cells, and the associated brackets specify the extremes of the observed length versus time data for all the cells tested. Figure 6 presents the cumulated data for aspiration length versus time for the three suction pressures with the 3.5-μm pipette at room temperature. This figure illustrates the functional similarity of the white cell response to aspiration at the elevated pressures of 500 and 700 dynes/cm². Likewise, Figures 7, 8, and 9 demonstrate the functional similarity again but for variations in pipette inner diameter and temperature at a suction pressure of 500 dynes/cm². This functional similarity can be represented by the following equation:

$$L/R_P = C_0 + C_1 \cdot f(t),$$

where capital C_1 is the coefficient of proportionality between the function of time, $f(t)$, and the length to pipette ratio, L/R_P. Furthermore, the coefficient C_1 contains essentially all of the dependence on suction pressure, pipette dimension, temperature; and the function $f(t)$ appears to be a power relation, t^n, where n is about 0.5. In contrast to the deformation response of the white cell at pressures above 500 dynes/cm², the response at lower pressures was indicative of a threshold below which the white cell would not flow into the pipette; this is evidenced by the data in Figure 10 for an aspiration pressure of 200 dynes/cm² with two different pipette

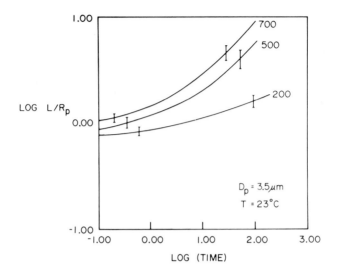

Fig. 6. Data for aspiration length versus time for three suction pressures with a 3.5-μm pipette at room temperature. Note the functional similarity of the granulocyte responses to aspiration at the elevated pressures of 500 and 700 dynes/cm^2. The curves represent the average of measurements on 5–10 cells and the associated bars give the extremes of the observed data for all the cells tested (taken from Evans and Kukan [1983]).

sizes. This threshold behaviour is explicitly demonstrated in Figures 11 and 12, which are plots of the aspiration length after 60 seconds duration versus the pipette suction pressure for different-sized pipettes.

The times to transition from the passive to active, motile cellular state observed in the deformation experiments are listed in Tables I and II (data taken from Evans and Kukan [1983]). A priori, it was anticipated that any such transition would be indicative of the contact between the white cell and the glass micropipette. However, this data showed some very interesting features. First of all, the activation time varied between 1 and 2 minutes at room temperature, with a mean of about 1.5 minutes. This activation only occurred when the cell had been deformed but not when it was simply held by the pipette below the threshold level for deformation. On the other hand, the activation time at 37°C varied over a range of 0 to 2 minutes with a mean of about 1 minute. The much larger variation in behaviour was characteristic of all aspects of the experiments at 37°C. Here, it appeared that contact with glass could have stimulated cell activity. It should be emphasized that there was no adhesion to the pipettes in any of these experiments performed in plasma. The cells always entered and exited the pipette freely without sticking. It is not certain whether the

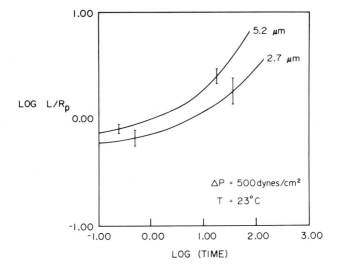

Fig. 7. Data for aspiration length versus time at a fixed pressure of 500 dynes/cm² with two different-sized pipettes of 2.7- and 5.2-micron diameters. The curves represent the average of measurements on 5–10 cells with the associated bars giving the extremes of the observed data for all the cells tested (taken from Evans and Kukan [1983]).

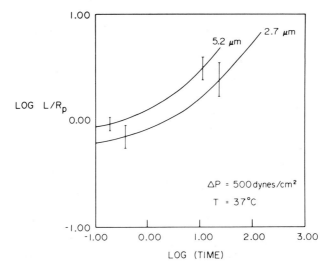

Fig. 8. Data for aspiration length versus time for a fixed suction pressure of 500 dynes/cm² at 37°C, again with two different-sized pipettes of 2.7 and 5.2 μm in diameter. The curves represent the average of measurements on 5–10 cells with the associated bars giving the extremes of the observed data for all the cells tested (taken from Evans and Kukan [1983]).

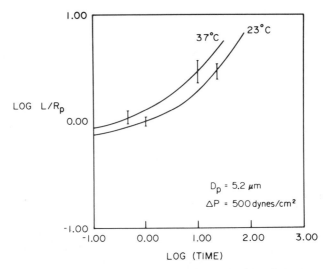

Fig. 9. Data for aspiration length versus time for a fixed suction pressure of 500 dynes/cm² and fixed pipette diameter of 5.2 μm and temperatures of 23°C and 37°C. The curves represent the average of measurements of 5–10 cells with the associated bars giving the extremes of the observed data for all the cells tested (taken from Evans and Kukan [1983]).

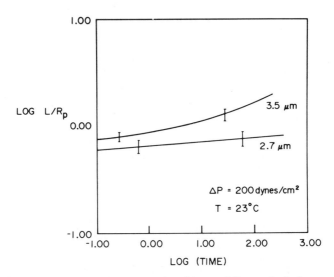

Fig. 10. Data for aspiration length versus time for two different-sized pipettes at a low aspiration pressure of 200 dynes/cm² and room temperature. These curves illustrate the threshold below which the white cell would not flow into the pipettes. The curves represent the average of measurements on 5–10 cells with the associated bars giving the extremes of the observed data for all the cells tested (taken from Evans and Kukan [1983]).

TABLE I. Time (sec) to Transition From Passive to Motile Cellular State in Deformation Test at 23°C

ΔP (dynes/cm^2)	$D_p = 2.7\ \mu m$	$D_p = 3.5\ \mu m$	$D_p = 5.2\ \mu m$
200	a	122 ± 74	100
500	102 ± 44	107 ± 60	b
700	115 ± 27	88 ± 34	b

[a]No transition observed.
[b]Totally aspirated into pipette with no transition observed.

TABLE II. Time (sec) to Transition From Passive to Motile Cellular State in Deformation Test at 37°C

ΔP (dynes/cm^2)	$D_p = 2.7\ \mu m$	$D_p = 3.5\ \mu m$	$D_p = 5.2\ \mu m$
200	68 ± 70	73 ± 56	a
500	64 ± 51	65 ± 46	a
700	74 ± 94	76 ± 39	a

[a]Totally aspirated into pipette with no transition observed.

contact with glass or the extension of the cell was responsible for the activation; however, these experiments indicate that deformation of the white cell can stimulate activity.

The results from the recovery after deformation observations were also interesting. First of all, the cell major dimension recovered in an essentially linear manner to that of a nearly perfect sphere with its original diameter. Hence, we simply quantitated the recovery process by measurement of the slope of the major cell dimension versus time. Tables III and IV (data taken from Evans and Kukan [1983]) list the reciprocals of these slopes as time constants versus pipette dimension and aspiration pressure at both temperatures. It appears from the data that the recovery time constant is essentially independent of aspiration pressure and pipette dimension but does depend on temperature. For example, the recovery time constants measured at 37°C are about twice as fast as those measured at room temperature. Again, the most impressive aspect of the recovery phase was that the cell recovered to an almost perfect sphere along a linear time course. This recovery feature was independent of extension and site of aspiration. These observations strongly indicate that the recovery is driven by a membrane tension and is opposed simply by a highly viscous but liquidlike interior.

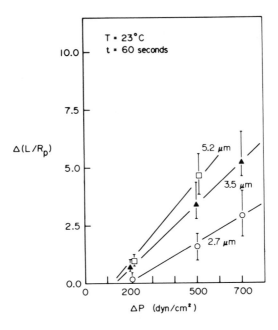

Fig. 11. Data for the aspiration length at a specific time of 60 seconds at room temperature versus the aspiration pressure for the three different-sized pipettes. The extrapolation to the abscissa clearly demonstrates the threshold below which the white cell would not flow into the pipette. The data are plotted as averages with bars which represent the extremes of the data for all cells tested (taken from Evans and Kukan [1983]).

TABLE III. Recovery Time Constant (sec) After Large Deformation at 23°C

ΔP (dynes/cm^2)	$D_p = 2.7\ \mu m$	$D_p = 3.5\ \mu m$	$D_p = 5.2\ \mu m$
200	a	30 ± 14	30
500	25 ± 9	22 ± 6	22 ± 6
700	32 ± 9	22 ± 8	22 ± 6

[a]No transition observed.

TABLE IV. Recovery Time Constant (sec) After Large Deformation at 37°C

ΔP (dynes/cm^2)	$D_p = 2.7\ \mu m$	$D_p = 3.5\ \mu m$	$D_p = 5.2\ \mu m$
200	11 ± 3	13 ± 5	13 ± 6
500	15 ± 9	13 ± 4	20 ± 12
700	12 ± 4	14 ± 5	15 ± 7

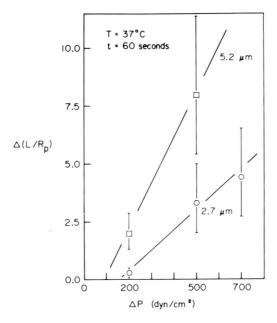

Fig. 12.   Data for aspiration length at a time of 60 seconds and an elevated temperature of 37°C versus pipette suction pressure for two pipette diameters. Again, the extrapolation to the abscissa demonstrates the threshold below which the white cell will not flow into the pipette (taken from Evans and Kukan [1983]).

We tested the capability of granulocytes to phagocytose zymosan particles at 23°C and 37°C in phosphate-buffered saline (PBS) and PBS plus 10% plasma. We found that the granulocytes would not adhere to the zymosan particles in PBS, and thus no phagocytosis occurred. On the other hand, with 10% plasma, the granulocyte first adhered to the particle and then proceeded to engulf the particle by a continuous spread of surface and cytoplasm around the particle. This material (for engulfment) emanated from a local region, as shown in Figure 5. The phagocytic process was considerably slower at 23°C than at 37°C. It is clear that the process involves direct surface recognition, since only one zymosan particle (and no other material in solution) was involved and that contact had to be formed before phagocytosis would proceed. Occasionally, adhesion would occur but was not followed by phagocytosis of the particle.

CONCLUSIONS

The passive deformation of granulocytes in the micropipette suction experiment appeared to be a continuous-flow process with no approach to

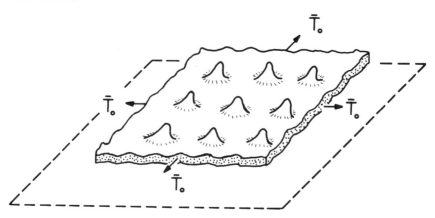

Fig. 13. Schematic illustration of the granulocyte membrane and subsurface material as a "contractile surface carpet" which possesses a membranelike tension that could arise from contractile elements tangent to the membrane surface. Surface contraction would cause membrane wrinkles and folds giving the carpet an appearance of a ruffled surface; the projected area of the carpet could be expanded until the "ruffles" were smoothed-out as illustrated by the dashed projection.

a static deformation limit. There was, however, an obvious threshold below which the cell would not deform and enter the micropipette. For suction pressures significantly above the threshold, granulocytes were continuously deformed with a remarkably similar functional dependence on time; the ratio of aspiration length to pipette radius increased essentially in proportion to time to about the one-half power. The coefficient of proportionality depended on suction pressure, pipette dimension, and temperature. It is interesting to note that this type of functional relationship does not exhibit the properties represented by the "standard visco elastic solid" model, ie, an instantaneous elastic response and an eventual approach to a static deformation limit. It was observed that the granulocytes always recovered to the spherical state after deformation, independent of the extent of deformation or the location where the cell was aspirated. This recovery behavior strongly indicates that the process is driven by a membranelike surface tension and opposed by viscous resistance of an interior liquid. Furthermore, the weak dependence on pipette dimension exhibited by the threshold pressure shown in Figures 10, 11, and 12 is consistent with the notion of an initial membrane tension. The concept that is being introduced here is that the granulocyte membrane and subsurface structure behave like a "contractile surface carpet," schematically illustrated in Figure 13. Such a carpet would exist in a ruffled state with a limit to its area expansivity given by the amount of material contained in the "ruffles," as illustrated in Figure 13. The membrane tension could arise from contractile elements tangent to the membrane surface; surface contraction would cause membrane wrinkles and folds to form until the surface forces are adequately opposed by stresses from the interior. Since the final shape is spherical, any internal stresses would appear to be uniform and characteristic of a liquidlike substance. Hence, this model for the passive state would imply that the granules, nucleus, and other discrete organelli are simply floating in the liquid cytoplasm. If we apply this model to the data shown in Figures 11 and 12, we derive values for the initial membrane tension of 1.3×10^{-2} dynes/cm at 23°C and 1.1×10^{-2} dynes/cm at 37°C. It would appear that temperature has little effect on this initial tension; however, because of the variance in the data at the elevated temperature, it is clear that some cells could have much lower values of initial tension. On the average, both the deformation data shown in Figures 11 and 12 and the recovery time constant data given in Tables II and IV indicate that temperature primarily lowers the effective viscosity of the cell interior. Additional evidence that the cell surface possesses an initial tension is provided by the observation that the cell recovery follows essentially a linear dependence on time as opposed to an exponential decay. The recovery dynamics represent the balance of elastic restoring forces against viscous dissipation in the interior; thus, if the elastic restoring forces were decreasing to zero in the approach to the final spherical state, the cell dimension would recover in an exponential manner. On the other hand, if the restoring forces did not diminish to zero, the cell would recover directly to the spherical state. Another fascinating feature of the recovery after deformation is that it appears to be faster than flow of the cell into the pipette. If we incorporate this observation into a contractile surface model, the implication is that the tension in the membrane surface increases with membrane extension and, thus, would exhibit a very rapid recovery from large extensions. Within the concept of this model, it is interesting to conjecture on mechanisms for pseudopod formation. First of all, formation of a long process (as shown in Fig. 14) would require local reduction in the passive contractile tension; this would permit expulsion of the cytoplasm into the process. In addition, there would have to be extensional deformation of the membrane surface (that is, local regions of the surface would extend from squares into long rectangles in contrast to the simple area expansion illustrated in Fig. 13). As such, there could also be a deviation between the membrane tension component along the process and the circumferential tension component (again in contrast to the "isotropic" tension illustrated in Fig. 13). This could be accomplished by alignment of subsurface contractile elements. Even though the "contractile membrane

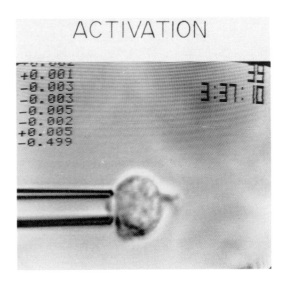

Fig. 14.   Video micrograph of a granulocyte which has formed a pseudopod.

carpet" model is speculative at this point, the threshold pressure data and recovery behaviour clearly support its proposition.

The observations of passive to active transition in the pipette suction experiment seemed to indicate that granulocytes may be stimulated by deformation, certainly at room temperature. However, the time to transition was independent of extension, and the cells were much more labile at 37°C, which make the data difficult to interpret.

ACKNOWLEDGMENTS

This work was supported in part by National Institutes of Health, US Public Health Service grant HL 26965. The author is grateful to Ms. Barbara Kukan who performed the micromechanical experiments.

REFERENCES

Bagge U (1976): Granulocyte rheology. Blood Cells 2:481–490.
Evans E, Kukan B (1983): Large deformation, recovery after deformation, and activation of granulocytes. Blood (submitted).
Lichtman MA, Kearney EA (1976): The filterability of normal and leukemic human leukocytes. Blood Cells 2:491–506.

Lichtman MA, Weed RI (1972): Alteration of the cell periphery during granulocyte maturation: Relationship to cell function. Blood 39:301–316.

Martin RR, Dewitz TS, McIntire LV (1979): Alterations in leukocyte structure and function due to mechanical trauma. In Hwang, Gross, Patel (eds): "Quantitative Cardiovascular Studies Clinical and Research Applications of Engineering Principles." Baltimore: University Park Press, pp 419–454.

Miller ME, Myers KA (1975): Cellular deformability of the human peripheral blood polymorphonuclear leukocyte: Method of study, normal variation and effects of physical and chemical alterations. Res J Reticuloendothel Soc 18:337–345.

Schmid-Schönbein GW, Shih YY, Chien S (1980): Morphometry of human leukocytes. Blood 56:866–875.

Schmid-Schönbein GW, Sung KLP, Tozeren H, Skalak R, Chien S (1981): Passive mechanical properties of human leukocytes. Biophys J 36:243–256.

II. WHITE CELL CYTOPLASMIC AND MEMBRANE PROPERTIES

White Cell Mechanics: Basic Science and
Clinical Aspects, pages 75–86

Molecular Architecture of the Cytoplasmic Matrix

Thomas D. Pollard

*Department of Cell Biology and Anatomy, Johns Hopkins Medical School,
Baltimore, Maryland 21205*

INTRODUCTION

My objective is to outline the present state of our knowledge concerning the molecular structure of the cytoplasmic matrix—that array of fibrous elements that is responsible for the gellike consistency of the cytoplasm and the physical integrity of the cell. Before 1970 much was known, at least qualitatively, about the physical properties of the cytoplasm, but there was essentially no information regarding the molecular basis for this behavior. Since then three types of cytoplasmic fibers (Fig. 1, Table I)—actin filaments, microtubules, and intermediate filaments—have been identified, purified, and characterized in some detail. Each fiber type is composed of a distinctive protein subunit and each has its own collection of accessory proteins. In the case of actin, there is now a fairly clear picture of how several classes of these proteins regulate actin polymerization and actin filament interactions. Comparable regulatory proteins probably exist for the other fiber types.

It is very likely that actin filaments, intermediate filaments, and microtubules are collectively responsible for the gellike properties of the cytoplasmic matrix. The physical stability of this network depends on connections between the fibrous elements that are maintained by specific proteins. In the case of actin, the best-understood example, actin filaments can bind to each other via specific cross-linking proteins; actin filaments can bind to microtubules via one or more of the microtubule-associated proteins; and actin filaments can bind to membranes. Comparable cross-links are likely for intermediate filaments and microtubules. The consequence is that the cell can form a continuous, cross-linked fiber network

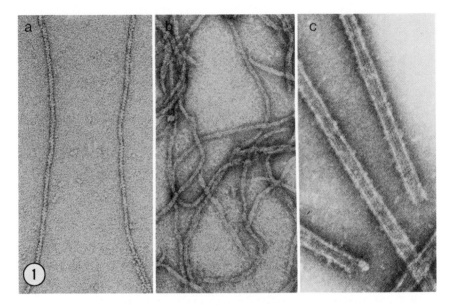

Fig. 1. Electron micrographs of (a) actin filaments, (b) keratin intermediate filaments, and (c) brain microtubules. Samples negatively stained with uranyl acetate.

extending from the plasma membrane throughout the cytoplasm. The term "cytoskeleton" is widely used to name this structural framework.

HISTORY

Since the earliest days of cytology in the 1840s [Dujardin, 1843], at least some biologists have understood that the cell is not a membrane-bounded bag containing soluble components in a saline solution. Rather, the cytoplasm has a gellike consistency with the following properties (reviewed by Frey-Wyssling [1948], Allen [1961], and Pollard [1976]): (1) the gel is found throughout the cytoplasmic matrix but can vary in consistency from place to place in the cell and also with time; (2) the gel is isotropic or only weakly birefringent; (3) the gel limits the amplitude of the Brownian movements of the cell organelles; (4) the gel forms by an endothermic process; (5) the gel is thixotropic; and (6) the gel appears to be contractile and likely to be involved with cellular movements.

These properties suggested to some that the gel consists of a random network of cross-linked filaments (Fig. 2). This interpretation has received widespread support from the biochemical and ultrastructural studies of the past 15 years. Some of these important facts were ignored during the

TABLE I. Three Classes of Cytoplasmic Fibers

Class	Protein subunit molecular weight	Distribution
1. Actin filaments (microfilaments)	Monomer 42,000	Universal in eukaryotes
2. Microtubules	Heterodimer 55,000 alpha-chain 50,000 beta-chain	Universal in eukaryotes
3. Intermediate filaments (IFs)		Metazoa
Keratin	40,000 to 68,000	Most epithelial cells
Vimentin	58,000	Mesenchymal cells
Desmin	55,000	Muscle
Neurofilaments	68,000 150,000 220,000	Neurons
Glial IFs	51,000	Glial cells

1950s and 1960s when the structural elements of the gel were not visible in early electron micrographs. Improved specimen preparation techniques for electron microscopy and a large body of biochemical work have clarified the situation. I will first summarize the fundamental properties of each fiber type and then discuss the evidence for interactions between the different fiber types and their association with membranes.

ACTIN

Actin is the most prominent protein in many cell types, often constituting 10–15% of total cellular protein (see reviews by Pollard and Weihing [1974] and Korn [1982]). The actin molecule has a molecular weight of 42,000 and is shaped like a pear. It is also the most widely distributed of the cellular fibers, being found in essentially all eukaryocytes, including both animal and plant cells. Consequently, I believe that it is *the* fundamental component of the cytoskeleton. Under physiological conditions actin self-assembles into filaments. Regrettably, the structure of these filaments is still not clear enough to be certain how the actin molecules are bound to each other.

The polymerization process consists of four steps (reviewed by Pollard and Craig [1982] and Korn [1982]): (1) Monomer activation is a first-order reaction that occurs when the molecule is exposed to Mg^{++} [Cooper et al, 1983]. This change in the actin molecule increases the rate of nucleation. (2) Nucleation is the formation of actin trimers, which can initiate the growth of an actin filament. These trimers are very unstable, and their formation limits the rate of the polymerization process. (3)

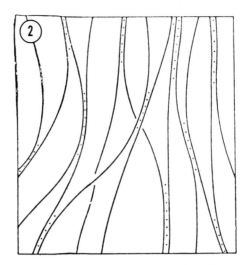

Fig. 2. Drawing of Frey-Wyssling's [1948] concept of the filamentous structure of the cytoplasmic matrix.

Elongation of actin filaments occurs very rapidly particularly at the "barbed end." The association of subunits with filament ends appears to be a diffusion-limited reaction that is faster by at least 1,000 than nucleation. (4) Actin filaments can also break in the middle and anneal end to end. At some point after an actin molecule is incorporated into a filament, the terminal phosphate is cleaved from the bound ATP, but the energy released is not required for polymerization.

In the cell, actin polymerization is controlled in at least three different ways by actin-binding proteins (Fig. 3) (reviewed by Craig and Pollard [1982] and Korn (1982)). One class of proteins, exemplified by profilin [Carlsson et al, 1977; Tobacman and Korn, 1982; Tseng and Pollard, 1982], binds to actin monomers, forming a complex that cannot polymerize. Such monomer-sequestering proteins can buffer the free actin monomer concentration in the cytoplasm. Perhaps even more important, profilin very strongly inhibits the nucleation reaction. This may prevent the spontaneous formation of free actin filaments in the cell.

A second class of regulatory proteins binds to the ends of the actin filaments, either the rapidly growing "barbed" end of the filaments or the slowly growing "pointed" ends of the filaments (Fig. 3) [Isenberg et al, 1980; Glenney et al, 1981]. The capping proteins that bind to the barbed end have been studied more extensively. These barbed-end capping proteins have multiple effects on actin polymerization. The most obvious is that they prevent the addition of subunits to the barbed end of the

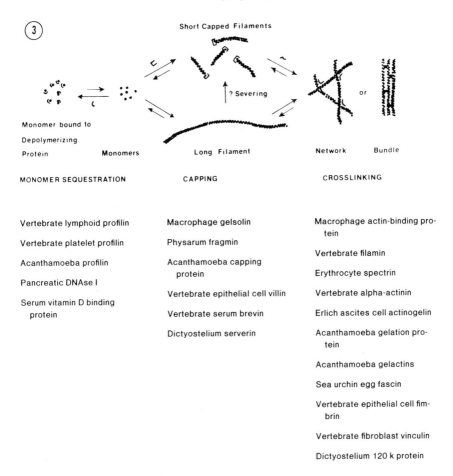

Fig. 3. Regulation of actin polymerization and cross-linking by three classes of actin-binding proteins. Monomer-sequestering proteins like profilin inhibit polymerization by binding to actin monomers. Capping proteins block one end of actin filaments and nucleate polymerization; some, like gelsolin, may sever preformed filaments. Cross-linking proteins attach filaments to each other to form networks or bundles.

filament. A second effect is their ability to nucleate actin polymerization, most likely by stabilizing actin dimers. Consequently, these proteins are in a position to specify the direction, rate, and site of actin polymerization in the cell. As one might expect, the capping proteins and profilin have antagonistic effects on actin nucleation: Together they may suppress spontaneous nucleation and specify the places that actin filaments form in the cell. Some, but not all, of the capping proteins require calcium for their activity (see, for example, Yin et al [1980]). Further, some of these capping proteins appear to have the ability to sever actin filaments [Yin et al, 1980; Hasegawa et al, 1980].

The third class of actin-binding proteins cross-links actin filaments, either in bundles or into random three-dimensional networks (Fig. 3). This class of proteins comes in many sizes and shapes. Some of these proteins, exemplified by macrophage actin-binding proteins [Hartwig and Stossel, 1981] are long, floppy molecules with molecular weights of ~500,000. A second group of actin cross-linking proteins is similar in size, shape, and subunit composition to the protein alpha-actinin that is found in the Z-lines of striated muscle (see, for example, Pollard [1981] and Burridge and Feramisco, [1981]). These proteins have a native molecular weight of about 180,000 and consist of two 90,000 molecular weight subunits. In at least one cell type, *Acanthamoeba,* there are actin cross-linking proteins that are much smaller, having molecular weights of about 30,000 [Maruta and Korn, 1977]. In addition to these proteins that are thought to be specific actin cross-linking proteins in the cell, a large number of basic macromolecules including polylysine, ribonuclease, lysozyme, and histones also can cross-link actin filaments [Griffith and Pollard, 1982a] filaments. These are thought to be nonspecific interactions. Some, but not all, of the specific cross-linking proteins have their activity inhibited by micromolar concentrations of calcium. The structures formed when these cross-linking proteins interact with actin filaments differ with the various proteins. In some cases, such as the alpha-actininlike cross-linking proteins, a three-dimensional random network of cross-linked filaments is formed. In other cases, such as the cross-linking protein villin from the microvilli of the intestinal brush border [Glenney et al, 1981], the actin filaments are aggregated into parallel bundles.

MICROTUBULES

The subunit of microtubules is a protein called tubulin (see review by Correia and Williams [1983]). It consists of two nonidentical, but related, 55,000 molecular weight polypeptides called alpha and beta tubulin. Like actin, tubulin dimers assemble into polymers by a nucleation/elongation mechanism. As in the case of actin polymerization, tubulin polymerization is accompanied by the hydrolysis of the bound nucleoside triphosphate, in this case GTP. The polymer is a cylinder about 25 nm in diameter, consisting of a helical arrangement of tubulin molecules. As in the case of actin filaments, the asymmetry of the subunits gives each filament a specific polarity. Again, like actin, the polymer grows much more rapidly at one end than the other.

When microtubule protein is isolated from brain or other tissues by cycles of polymerization and depolymerization, microtubule-associated proteins (MAPs) copurify with the tubulin subunits. In brain there are two major classes of MAPs: high molecular weight MAPs and tau proteins. Both of these MAP protein classes bind fairly strongly to microtubules. The high molecular weight MAPs are large enough to visualize by electron microscopy and are seen as radial projections from the tubule surface. Both the high molecular weight MAPs and tau stabilize microtubules by decreasing the rate that tubule subunits dissociate from the ends of the polymer. As of this time, no molecules comparable to the actin monomer-binding protein profilin or comparable to the actin-capping proteins have been identified in association with tubulin. However, it is strongly suspected that such proteins exist. Likewise, no clear-cut examples of microtubule cross-linking proteins have been identified in association with cytoplasmic microtubules. There are, of course, examples of tubule cross-linking proteins in highly ordered microtubule structures such as the axonemes of cilia. Consequently, microtubule cross-linking proteins may also be found in the cytoplasm.

INTERMEDIATE FILAMENTS

The term "intermediate filament" is used to describe a widely distributed but heterogeneous class of filaments about 10 nm wide (Table I) (reviewed by Osborn et al [1981]). The best-characterized intermediate filament protein is keratin from epidermis and other epithelial tissues. Intermediate filaments composed of slightly different protein subunits have been found in many other cell types, and it is now recognized that there are intermediate filaments specific for muscle cells, for epithelial cells, for glial cells in the brain, for neurons, and for cells of mesenchymal origin (Table I). All of these intermediate filament protein subunits can be distinguished by immunological criteria, but recent structural studies have shown that they probably all have a similar fundamental structure. All of these proteins are very rich in alpha helical regions that have long been recognized in keratin. At the present time it seems likely that the subunits of the intermediate filaments consist of three polypeptide chains and that there are three or four strands of these subunits in each 10-nm wide filament.

Compared with actin and tubulin, little is known about the associated proteins for intermediate filaments. However, one protein called filaggrin [Steinert et al, 1981] has been identified and shown to aggregate intermediate filaments into bundles.

Compared with actin filaments and microtubules, the intermediate filaments are very stable. For example, they resist extraction from cells by high concentrations of salt and detergents. Another difference is that intermediate filaments may be limited in their distribution to metazoan cells. Thus, they cannot be considered to be universal components of the cytoskeleton like actin filaments and microtubules.

INTERACTIONS BETWEEN THE THREE FIBER SYSTEMS

The interaction between actin filaments and microtubules has been demonstrated by biochemical experiments with purified proteins. Mixtures of purified actin with microtubules consisting of tubulin and MAPs form a complex that has a viscosity considerably higher than the viscosity of the sum of the components (Griffith and Pollard, 1978, 1982b]. This interaction depends on the presence of the MAP proteins. Both the high molecular weight MAPs and tau appear to be involved, because both high molecular weight MAPs and tau are capable of cross-linking purified actin filaments into a three-dimensional network with high viscosity. Further, while mixtures of purified actin filaments and pure tubulin polymers do not have a high viscosity, the complex can be reconstituted from mixtures of purified actin, tubulin and MAPs.

The interaction of actin filaments and microtubules appears to be regulated, at least in part, by phosphorylation of MAPs [Selden and Pollard, 1983]. Phosphorylation of both high molecular weight MAPs and tau inhibits their actin cross-linking activity.

In a similar way, mixtures of microtubule protein and intermediate filaments from brain can form a high-viscosity complex providing ATP is present [Runge et al, 1980]. Little is known about the factors that regulate the interactions between these two fiber types. Further, no work has yet been reported on interactions of actin filaments with intermediate filaments.

Since actin filaments are the only one of these three fiber types found universally in cells, it seems likely to me that they must be the fundamental structural component of the cytoplasmic matrix in most cells. The presence of microtubules and intermediate filaments in many cells and the ability of at least the microtubules to interact with actin filaments suggests that microtubules and intermediate filaments may complement, and interact with, the actin filament system in maintaining the structure of the cytoplasmic matrix.

STRUCTURE OF THE CYTOPLASMIC MATRIX

Visualization of the cytoplasmic matrix fibers was hampered for years by inadequate electron microscopic techniques. Initially, there was difficulty preserving the three fiber types for electron microscopy. Modern fixation procedures have overcome this difficulty to a large extent.

However, the traditional method of cutting thin sections through cells is not particularly advantageous for seeing the cytoplasmic matrix fiber system. Fortunately, there are two new techniques that allow one to see the structure of the cytoplasmic matrix in three dimensions. The first, championed by Keith Porter and his colleagues [Wolosewick and Porter, 1979], involves fixing cells and preparing them for electron microscopy by critical-point drying. The dried cells are then viewed in a high-voltage electron microscope. The cytoplasmic matrix is seen to be full of a branching and anastomosing network of fine filaments that have been termed microtrabeculae. These microtrabeculae appear to interact both with intermediate filaments and with microtubules. The molecular components of the microtrabeculae are not known with certainty, however, it seems likely that actin is a major component. The second method involves rapidly freezing cells using techniques perfected by John Heuser and his colleagues [Heuser and Kirshner, 1980]. The frozen cells are fractured in a freeze-fracture machine and the volatile components of the cytoplasm removed by sublimation. The remaining material is then contrasted by shadowing with platinum and carbon and the replica is viewed in a conventional transmission electron microscope. This provides a dramatic three-dimensional view of the three fiber types in the cytoplasm. The detail preserved in these replicas is spectacular, because it is even possible to distinguish individual actin molecules and tubulin molecules in their respective filament types. The cytoplasmic matrix is observed to be filled with a dense network of actin filaments in addition to intermediate filaments and microtubules.

Although both of these methods allow one to see the three-dimensional arrangement of the fiber types in the cytoplasm, neither method is capable of identifying which cytoplasmic fibers are physically cross-linked to each other. This is demonstrated dramatically by model experiments with purified actin [unpublished work of T. D. Pollard and J. Heuser]. We found that the arrangement of filaments in a solution of purified actin having an apparent viscosity of <100 centipoise (cp) is identical to the arrangement of actin filaments in gels formed from actin filaments cross-linked with an alpha-actininlike protein having an apparent viscosity >10,000 cp. In both cases there are contacts between the randomly disposed actin filaments. However, in the actin filament solution the contacts must be very transient or weak. In the gels some of the contacts must be stabilized by the actin cross-linking proteins, although the actual molecular details are not visible in the micrographs.

INTERACTION OF CYTOPLASMIC FIBERS WITH MEMBRANES

There are three well-documented examples of actin filament interactions with membranes, but the only system that is understood in any molecular detail is found in the red blood cell. There it is possible to trace the molecular interactions from short actin oligomers to the protein spectrin, from spectrin to the protein ankyrin, and from ankyrin to the intrinsic membrane protein named band 3 (reviewed by Branton et al [1981]). Thus, the bond between the actin filament and the membrane involves at least two peripheral membrane proteins, ankyrin and spectrin.

In the microvilli of the intestinal brush border (reviewed by Mooseker and Howe [1982]) there is excellent morphological evidence that actin filaments interact with the membranes at their barbed ends, at the tip of the microvilli, and laterally along the sides of the microvilli. At the tip there is some dense material located at the site where the barbed end of the actin filaments appears to contact the membrane. Along the side of the actin filament bundle in the microvilli there are radial links extending between the actin filament bundle and the surrounding membrane. Nothing is known about the molecular components of the dense material at the tip of the microvilli, but there is evidence that a protein with a subunit molecular weight of 110,000 forms at least in part the radial link between the actin filaments and the membrane. The intrinsic membrane proteins bound to the radial link have not been identified.

In many other cell types there are both morphological observations and biochemical evidence, obtained by cell fractionation and plasma membrane purification, that actin filaments are bound to the plasma membrane. As in the case of the brush border, the barbed ends of the actin filaments appear to be associated with the membrane. This, of course, suggests that the plasma membrane of nonmuscle cells may function as an actin filament-anchoring site just as the Z-line does in striated muscle. In both cases the barbed end of the actin filament is attached. In muscle the tension generated by interaction of these actin flaments with myosin pulls on the attachment site, and we would expect the same to be true in nonmuscle cells.

Intermediate filaments bind to the plasma membrane at specializations called desmosomes (reviewed by Fawcett [1981]). These are punctate specializations of the plasma membrane that bind the membrane to a desmosome of an adjacent cell or to the basement membrane underlying an epithelial cell. On the cytoplasmic surface of the desmosome intermediate filaments form looping lateral contacts with electron dense, but currently uncharacterized material on the cytoplasmic face of the membrane.

In contrast to intermediate filaments and actin filaments, little evidence is available to suggest that microtubules interact with the plasma membrane except at the tip of cilia and flagella. However, there is both morphological and biochemical evidence that membranes of organelles can be bound to microtubules in the cytoplasm.

The associations of both actin filaments and intermediate filaments with the plasma membrane are undoubtedly essential for the physical integrity of the cell. These bonds allow tension generated in the cytoplasmic actin-myosin system to be transmitted to the plasma membrane and from the plasma membrane to the substrate or to adjacent cells. Likewise, these contacts between the cytoplasmic matrix fibers and the plasma membrane allow stresses applied to the surface of the cell to be distributed throughout the cytoplasm. In some cases where the stress is particularly high as on the free surface of the endothelial cells lining the heart and the large blood vessels, the actin filaments can be gathered into large bundles called stress fibers which probably reinforce the cell [Wong et al, 1983]. Similarly, the concentration of intermediate filaments and the number of desmosomes is nowhere higher than in the epidermis where the epithelium is under constant physical abuse.

REFERENCES

Allen RD (1961): Amoeboid movement. In: "The Cell." New York: Academic Press, Inc., Vol 2, pp 135–216.

Branton D, Cohen CM, Tyler J (1981): Interaction of cytoskeletal proteins on the human erythrocyte membrane. Cell 24:24–32.

Burridge K, Feramisco JR (1981): Non-muscle alpha-actinins are calcium sensitive actin-binding proteins. Nature 294:565–567.

Carlsson L, Nystrom LE, Sundkvisk I, Markey F, Lindberg U (1977): Actin polymerizability is influenced by profilin, a low molecular weight protein in non-muscle cells. J Mol Biol 115:465–483.

Cooper JA, Buhle EL Jr, Walker SB, Tsong TY, Pollard TD (1983): Kinetic evidence for a monomer activation step in actin polymerization. Biochemistry 22:2193–2202.

Correia JJ, Williams RC (1983): Mechanisms of assembly and disassembly of microtubules. Annu Rev Biophys Bioeng 12:211–235.

Craig SW, Pollard TD (1982): Actin-binding proteins. Trends Biochem Sci 7:88–92.

Dujardin F (1843): Recherches sur les organismes inferieurs. Ann Sci Natl Zool 4:343.

Fawcett DW (1981): The Cell. Philadelphia: WB Saunders Co, pp 156–167.

Frey-Wyssling A (1948): Submicroscopic Morphology of Protoplasm and Its derivatives." New York: Elsevier Publishing Co., Inc.

Glenney JR, Kaulfus P, Weber K (1981): F-actin assembly modulated by villin: Ca^{++}-dependent nucleation and capping of the barbed end. Cell 24:471–480.

Griffith LM, Pollard TD (1978): Evidence for actin filament-microtubule interaction mediated by microtubule-associated proteins. J Cell Biol 78:958–965.

Griffith LM, Pollard TD (1982a): Crosslinking of actin filament networks by self-association and actin binding macromolecules. J Biol Chem 257:9135–9142.

Griffith LM, Pollard TD (1982b): The interaction of actin filaments with microtubules and microtubule associated proteins. J Biol Chem 257:9143–9151.

Hartwig JH, Stossel TP (1981): Structure of macrophage actin-binding protein molecules in solution and interacting with actin filaments. J Mol Biol 145:563–581.

Hasegawa T, Takahashi S, Hayashi H, Hatano S (1980): Fragmin: A calcium ion sensitive regulatory factor on the formation of actin filaments. Biochemistry 19:2677–2683.

Heuser JE, Kirschner MW (1980): Filament organization revealed in platinum replicas of freeze-dried cytoskeletons. J Cell Biol 86:212–234.

Isenberg GH, Aebi U, Pollard TD (1980): An actin binding protein from Acanthamoeba regulates actin filament polymerization and interactions. Nature 288:455–459.

Korn ED (1982): Actin polymerization and its regulation by proteins from non-muscle cells. Physiol Rev 62:672–737.

Maruta H, Korn ED (1977): Purification from Acanthamoeba castellanii of proteins that induce gelation and syneresis of F-actin. J Biol Chem 252:399–402.

Mooseker MS, Howe CL (1982): The brush border of intestinal epithelium. A model system for analysis of cell surface architecture and motility. In Wilson L (ed): "Methods and Perspectives in Cell Biology." New York: Academic Press, Vol 25, pp 143–174.

Osborn M, Geisler N, Shaw G, Sharp G, Weber K (1981): Intermediate filaments. Cold Spring Harbor Symp Quant Biol 46:413–429.

Pollard TD (1976): Cytoskeletal functions of cytoplasmic contractile proteins. J Supramol Struct 5:327–334.

Pollard TD (1981): Purification of a calcium-sensitive actin gelation protein from Acanthamoeba. J Biol Chem 256:7666–7670.

Pollard TD, Craig SW (1982): Mechanism of actin polymerization. Trends Biochem Sci 7:55–58.

Pollard TD, Weihing RR (1974): Actin and myosin and cell motility. CRC Crit Rev Biochem 2:1–65.

Runge MS, Laue TM, Yphantis DA, Lifsics MR, Saito A, Altin M, Reinke K, Williams RC Jr (1981): ATP-induced formation of an associated complex between microtubules and neurofilaments. Proc Natl Acad Sci USA 78:1431–1435.

Selden SC, Pollard TD (1983): Phosphorylation of microtubule associated proteins regulates their interaction with actin filaments. J Biol Chem 258: 7064–7071.

Steinert PM, Cantieri JS, Teller DC, Longsdale-Eccles JD, Dale BA (1981): Characterization of a class of cationic proteins that specifically interact with intermediate filaments. Proc Natl Acad Sci USA 78:4097–4101.

Tobacmann LS, Korn ED (1982): The regulation of actin polymerizatn and the inhibition of monomeric actin ATPase activity by Acanthamoeba profilin. J Biol Chem 257:4166–4170.

Tseng P, Pollard TD (1982): Mechanism of action of Acanthamoeba profilin. Demonstration of actin species specificity and regulation by micromolar concentrations of $MgCl_2$. J Cell Biol 94:213–218.

Wolosewick JJ, Porter KR (1979): Micotrabecular lattice of the cytoplasmic ground substance. Artifact or reality. J Cell Biol 82:114–139.

Wong AJ, Pollard TD, Herman M (1983): Actin filament stress fibers in vascular endothelial-cells in vivo. Science 219:867–969.

Yin HL, Zaner KS, Stossel TP (1980): Ca^{++} control of actin gelation. Interaction of gelsolin with actin filaments and regulation of actin gelation. J Biol Chem 255:9494–9500.

White Cell Mechanics: Basic Science and
Clinical Aspects, pages 87–94

Polyphosphoinositides as Regulators of Membrane Skeletal Stability

Michael P. Sheetz, Wen-Pin Wang, and Donald L. Kreutzer

*Department of Physiology (M.P.S.) and Pathology (W.-P.W., D.L.K.),
University of Connecticut Health Center, Farmington, Connecticut 06032*

INTRODUCTION

The interface between the cell cytoplasm and the plasma membrane is critical to many cellular functions, since many external stimuli and cytoskeletal changes are expressed there. It is no surprise, therefore, that a specialized portion of the cell cytoskeleton exists at that interface. This region has been described as the cell cortex or, as we will refer to it here, the membrane skeleton. This membrane skeleton has been studied in detail in the human erythrocyte where the membrane skeleton can be isolated readily without cytoplasmic vesicles, microtubules, intermediate filaments, and many other cytoskeletal elements. Many of the association constants have been determined for erythrocyte skeletal proteins and the detailed architecture of the membrane and associated skeleton is understood [Branton et al, 1981; Palek and Lux, 1983]. Although the erythrocyte membrane is necessarily highly specialized, the major proteins of the membrane skeleton, spectrin, actin, and band 4.1, have analogs in every cell type tested [Glenney et al, 1982; Bennett et al, 1982; Goodman et al, 1981]. It is reasonable, therefore, to propose that similar functions may be served by other membrane skeletons and that similar control mechanisms may be used.

The membrane skeleton of the erythrocyte has a major role in determining cell deformability, cell shape, and glycoprotein lateral diffusion and in preventing membrane vesiculation [Sheetz, 1983]. However, in leukocytes cell deformability and shape are determined primarily by elements of the cytoskeleton. Glycoprotein lateral diffusion rates most likely provide the best measure of the properties of the membrane skeleton. The method of fluorescence recovery after photobleaching (FRAP) has been shown to provide valid measurements of the rates of diffusion of membrane glycoproteins in many systems [Koppel, 1982]. In leukocytes as in most other cells the glycoproteins have rapid- and slow-diffusing components which are designated the mobile and immobile components, respectively. The immobile fraction corresponds to the fraction of molecules which are attached to the membrane skeleton as in the case of the erythrocyte anion channel (band 3) [Sheetz, 1983]. Although this correlation has only been tested in a couple instances, it is generally accepted. It has been more difficult to explain why the "mobile" proteins in membranes diffuse at rates 10- to 100-fold lower than those of the same proteins in pure lipid bilayers or skeleton-free membranes. There are two major possibilities: The first is that the proteins are in equilibrium with binding sites on the skeleton [Koppel, 1981], and the second is that the membrane skeleton forms a matrix which restricts glycoprotein diffusion over long but not short distances [Koppel et al, 1981]. The binding-and-release model cannot explain the anisotropic diffusion observed in cultured cells [Smith et al, 1979] or the rapid rotational diffusion of membrane glycoproteins whose lateral diffusion is greatly restricted [Nigg and Cherry, 1980]. Those observations, however, are consistent with the matrix model. Also, a number of studies have shown that destabilizing the membrane skeleton increases [Sheetz et al, 1982; Schindler, et al, 1980], whereas stabilizing decreases [Smith and Palek, 1982], the diffusion rates. The correlation between skeleton stability and diffusion rate is predicted by the matrix model, since the matrix must break apart for proteins to pass through it. Under conditions where the skeletal attachment to the membrane surface is unaltered, changes in lateral diffusion rates can provide a good measure of the relative stability of the membrane matrix.

Control of the organization of the membrane matrix is at present poorly understood in cells other than the erythrocyte. Ca^{++} and pH have been invoked as modulators of membrane skeletal as well as cytoskeletal stability. The proximity of the membrane skeleton to the lipid bilayer affords the possibility that a lipid or lipid-associated compound could modify membrane skeletal stability without affecting the cytoskeleton directly. In the erythrocyte membrane skeleton such an agent has been found in triphosphoinositide (TPI). We have demonstrated that polyanions like TPI will decrease the affinity of spectrin for actin in vitro [Sheetz and Casaly, 1980, 1981]. In addition we have demonstrated that TPI will selectively increase the lateral diffusion rate of band 3 [Sheetz et al, 1982]. Because TPI may also affect the binding of spectrinlike proteins to actin in other cells, it may be a general modulator of membrane skeletal stability. To test this hypothesis, we have begun to study the effects of agents which modify TPI metabolism on cell function. One such agent is Trental, a methyl xanthine derivative, which is known to increase tissue perfusion without causing significant vasodilation. Our findings are preliminary but support the hypothesis that increased TPI levels caused by Trental may destabilize the membrane skeleton and increase cellular motile activity.

METHODS AND MATERIALS

Leukocyte Chemotaxis

Analysis of leukocyte migration was done utilizing the standard Boyden chamber system as previously described [Kreutzer et al, 1978]. Glycogen-induced neutrophils were isolated from the rabbit peritoneum as previously described and utilized for all leukocyte-migration and enzyme-release assays [Kreutzer et al, 1978]. The synthetic chemotactic peptide formyl-methinyl-leucyl-phenylalanine (f-met-leu-phe) was utilized as a chemotatic and enzyme releasing agent for these studies to determine the effect of Trental on leukocyte migration and chemotaxis. Varying concentrations of Trental were placed in the lower compartment of a Boyden chamber in the presence or absence of a constant concentration of f-met-leu-phe ($3ED_{50}$) which gives maximal chemotactic response.

Polyphosphoinositide Levels

The levels of polyphosphoinositides (di- and tripolyphosphoinositide) were measured using cells labeled to equilibrium (20 hr) with $^{32}P_i$ in standard Hank's medium [Keutzer et al, 1978]. Lipids were separated by the procedure of Schacht [1978], and erythrocyte membranes were added to the extraction to provide carrier lipids.

RESULTS

Chemotaxis

To test for effects of the drug Trental on cellular motile activities, a standard Boyden chamber analysis was employed in the presence and absence of the chemoattractant, f-met-leu-phe. Movement of the cells from one surface of the filter to the other was increased by low concentrations of the drug and inhibited at high concentrations (Fig. 1). The increased leukocyte migration through the filters was observed both in the presence and absence of the chemoattractant. When coupled with prelimi-

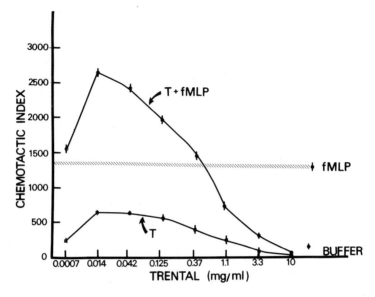

Fig. 1. The chemotactic index is plotted versus Trental concentration for rabbit neutrophils in the presence (O) and absence (□) of f-met-leu-phe.

nary studies of the effect of Trental on chemotaxis in a Zigmond-Hersch analysis these findings suggest that the low concentrations of the drug generally increase cellular motility without specifically being intrinsically chemotactic.

Polyphosphoinositide Levels

We wondered because of the observed effects of polyphosphoinositides on erythrocyte membrane skeletons whether or not changes in their levels occurred in parallel with changes in motile functions. Since the number of neutrophils was insufficient for a direct determination of the amount of phosphate in polyphosphoinositides, the cells were labeled with $^{32}P_i$ to equilibrium by preincubating them for 20 hr at 37°C. The ^{32}P contents of the diphospho- and triphosphoinositide fractions were increased by as much as eightfold in the rabbit peritoneal neutrophils upon treatment with 0.02 mg/ml of Trental, whereas little or no change was found at 2 mg/ml of Trental (Table I). This dose dependency parallels that of the chemotaxis and migration through the nucleopore filters in that significant stimulation occurs at the low concentration which is not found at high concentrations of the drug.

Controls were performed to determine that lipid extractions were unaltered by these experimental conditions. Further, the ^{32}P label in the

TABLE I. Effect of Trental on Rabbit Neutrophil Polyphosphoinositide Levels

Trental concentration	f-met-leu-phe	Diphosphoinositide[a]	Triphosphoinositide[a]
0	0	1 ± .2	1 ± .25
0.02 mg/ml	0	5.5 ± 1.5	5.1 ± 1.2
2.0 mg/ml	0	1.1 ± .3	0.9 ± .3
0	+	1 ± .2	1 ± .3
0.02 mg/ml	+	8.3 ± 2.4	8.3 ± 2.1
2.0 mg/ml	+	1.05 ± .3	.8 ± .3

[a]Levels of diphosphoinositide (DPI) and triphosphoinositide (TPI) are expressed in terms of the counts in DPI and TPI relative to the counts in those fractions in the control group of cells. The counts in the lipid extract which did not bind to the neomycin column, ie, lipids other than DPI and TPI, were not significantly altered in any samples. All points represent the average of six determinations.

flow-through fractions which represents other lipid fractions showed no significant changes with Trental and cannot, therefore, explain the large changes in the diphosphoinositide (DPI) and TPI fractions.

DISCUSSION

We would like to understand the role of the membrane skeleton in chemotaxis and other cellular motile phenomena. At present, however, we know very little of the mechanisms by which such processes occur. It is useful, therefore, at this stage to formulate hypotheses if only to disprove them later. It is suggested that the skeleton is restrictive of cellular motile and exocytic activity and that raising TPI levels in the cell could dissociate the skeleton and lessen that restriction. On the in vivo level the increase in motile activity may enable the cells to pass through capillaries more easily.

At the molecular level the effects of TPI are poorly understood. It is the most rapidly metabolized lipid in the plasma membrane and is modulated by a variety of hormones (see review by Hawthorne and Kai [1980]). Previously suggested functions for the lipid in Ca^{++} binding and transport have not borne up [Hawthorne, 1982]. In addition it has been suggested that TPI and DPI are reservoirs of arachidonic acid, but this hardly explains their rapid turnover. More recent studies have demonstrated that TPI could function to control membrane skeletal integrity and the formation of membrane pores by diptheria toxin [Donovan et al, 1982]. The latter is a specialized function which may in the future prove to have general correlates, but it is the former which is the focus of this paper. TPI is an ideal modulator of membrane skeletal integrity because it is restricted to the plasma membrane. Unlike Ca^{++} which will diffuse into the cytoplasm and modulate a variety of activities, TPI is largely resident in the plasma membrane. This occurs because the kinases which produce it are plasma membrane enzymes [Phillips, 1973; Schneider and Kirschner, 1970], whereas the phosphatase is cytoplasmic [Nijjan and Hawthorne, 1977] and would degrade any TPi which leaves the plasma membrane. In rabbit polymorphonuclear leukocytes (PMN) the increase in TPI levels cause by Trental was particularly dramatic at the optimal Trental concentration for cellular motility. A four- to eightfold increase in TPI and DPI levels is beyond the level which can be explained as a result of turnover of the diester phosphate of the PI precursor of TPI and DPI. Rather we suggest that there is a dramatic increase in the actual level of TPI and DPI in cells under conditions where cell motility is markedly altered.

Is there a causal relationship between the rise in TPI and DPI and the increase in cellular motility and enzyme release? These findings when coupled with observations in a variety of systems strongly implicate TPI and DPI as important components in motile processes. In human PMN there are also parallel increases in TPI, DPI, and motility. In platelet activation and in sea urchin egg fertilization [Turner, Jaff, and Sheetz, unpublished results] one of the earliest events is an increase in TPI and DPI. These facts coupled with the known action of these compounds in dissociating the erythrocyte membrane skeleton lead us to suggest that the rise in TPI and DPI destabilizes the membrane skeletons in these cells which facilitates the shape changes which occur. At this time many alternative hypotheses are tenable, but we now have a framework upon which to design further studies.

On the level of the tissue in vivo, there are dramatic increases in perfusion caused by Trental which are not mimicked by other methylxanthines. In related studies we have observed that human PMN show increased motility with Trental but not with caffeine or theophylline. Thus there is a correlation between increased white cell motility and increased tissue perfusion. There may be similar effects of Trental on capillary endothelial cells, but in any case the explanation of membrane skeletal destabilization caused by increased TPi and DPI levels is consistent with increased perfusion.

At present the role of the membrane skeleton in cellular mechanical properties is ill-defined. We must extrapolate from studies of the erythrocyte to explain how changes in cellular mechanical properties may be related to changes in the membrane skeleton. The correlations demonstrated here do indicate that important events are occurring at the interface between the plasma membrane and the cytoplasm wherein lies the membrane skeleton. Indeed it is most logical to point to that interface as the site of transduction of external stimuli into changes in cellular mechanical properties.

SUMMARY

The membrane skeleton is a two-dimensional complex of actin, spectrinlike, and associated proteins which lies on the cytoplasmic face of most plasma membranes. Components of this complex are believed to control the lateral mobility of integral membrane proteins as well as influence cell shape and motility. In earlier studies we observed that the addition of polyphosphorylated inositol lipids could increased membrane skeleton dissociation. In preliminary studies of leukocyte function we have observed that increased chemotaxis with Trental addition is correlated with increased polyphosphoinositide levels. Consequently, we suggest that polyphosphoinositides contribute to, if not are requisite for, cellular mobility.

ACKNOWLEDGMENTS

Dr. Sheetz is an Established Investigator of the American Heart Association. Dr. Kreutzer is an Established Investigator of the American Heart Association. This work was supported by a grant from Hoechst Roussel Pharmaceuticals.

REFERENCES

Bennett V, David J, Fowler WE (1982): Immunoreactive forms of erythrocyte spectrin and ankyrin in brain. Philos Trans R Soc Lond [Biol] 299:301–312.
Branton D, Cohen LM, Tyler J (1981): Interactions of cytoskeletal proteins on the human erythrocyte membrane. Cell 24:24–32.
Donovan JJ, Simon MI, Montal M (1982): Insertion of diptheria toxin into and across membranes: Role of phosphoinositide asymmetry. Nature 298:669–672.
Glenney JR, Glenney P, Weber K (1982): Erythroid spectrin, brain fodrin and intestinal brush border proteins (TW-260/240) are related molecules containing a common calmodulin-binding subunit bound to a variant cell type-specific subunit. Proc Natl Acad Sci USA 79:4002–4005.
Goodman S, Zagon IS, Kulikowski R (1981): Identification of a spectrin-like protein in nonerythroid cells. Proc Natl Acad Sci USA 78:7570–7574.
Hawthorne JN (1982): Is phosphatidylinositol now out of the calcium gate? Nature 295:281–282.
Hawthorne JN, Kai M (1980): Metabolism of phosphoinositides. In Lajtha (ed): "Handbook of Neurochemistry." New York: Plenum, pp 491–508.
Koppel DE (1981): Association dynamics and lateral transport in biological membranes. J Supramol Struct Cell Biochem 17:61–68.

Koppel DE (1982): Measurement of membrane protein lateral mobility. In Hesketh et al (eds): "Techniques in Lipid and Membrane Biochemistry—Part II." New York: Elsevier, pp 1–39.

Koppel DE, Sheetz MP, Schindler M (1981): Matrix control of protein diffusion in biological membranes. Proc Natl Acad Sci USA 78:3576–3580.

Kreutzer DL, O'Flaherty JT, Orr W, Showell HJ, Ward PA, Becker EL (1978): Quantitative comparisons of various biological responses of neutrophils to different active and inactive chemotactic factors. Immunopharmacology 1:39–47.

Nigg EA, Cherry RJ (1980): Anchorage of a band 3 population at the erythrocyte cytoplasmic membrane surface: Protein rotational diffusion measurements. Proc Natl Acad Sci USA 77:4702–4706.

Nijjan MS, Hawthorne JN (1977): Purification and properties of polyphosphoinositide phosphomonoesterase from rat brain. Biochim Biophys Acta 480:390–402.

Palek J, Lux SE (1983): Erythrocyte membrane skeleton in normal and disease states. Semin Hematol 20:189–224.

Phillips JH (1973): Phosphatidylinositol kinase. Biochem J 136:579–587.

Schindler M, Koppel DE, Sheetz MP (1980): Modulation of membrane protein lateral mobility by polyphosphates and polyamines. Proc Natl Acad Sci USA 77:1457–1461.

Schneider RP, Kirschner LB (1970): Di- and triphosphoinositide metabolism in swine erythrocyte membranes. Biochim Biophys Acta 202:283–294.

Schacht J (1978): Purification of polyphosphoinositides by chromatography on immobilized neomycin. J Lipid Res 19:1063–1067.

Sheetz MP (1983): Membrane skeletal dynamics: Role in modulation of red cell deformability, mobility of transmembrane proteins and shape. Semin Hematol 20:175–188.

Sheetz MP, Casaly J (1980): Diphosphoglycerate and ATP dissociate erythrocyte membrane skeletons. J Biol Chem 255:9955–9960.

Sheetz MP, Casaly J (1981): Phosphate metabolite regulation of spectrin-actin interactions. 41(Suppl 156):113–118.

Sheetz MP, Febbroriello P, Koppel D (1982): Triphosphoinositide increases glycoprotein lateral mobility in erythrocyte membranes. Nature 296:91–93.

Smith BA, Clark WR, McConnell HM (1979): Anisotropic molecular motion on cell surfaces. Proc Natl Acad Sci USA 76:5641–5644.

Smith DK, Palek J (1982): Modulation of lateral mobility of band 3 in the red cell membranes by oxidative cross-linking of spectrin. Nature 297:424–425.

White Cell Mechanics: Basic Science and
Clinical Aspects, pages 95–110
© 1984 Alan R. Liss, Inc., 150 Fifth Avenue, New York, NY 10011

Modulation of Tubulin Tyrosinolation in Human Polymorphonuclear Leukocytes (PMN)

Jayasree Nath and John I. Gallin
Bacterial Diseases Section, Laboratory of Clinical Investigation, National Institute of Allergy and Infectious Diseases, National Institutes of Health, Bethesda, Maryland 20205

INTRODUCTION

In recent years, studies from many laboratories have indicated significant alterations in the cytoskeletal organization of neutrophils (PMN), coincident with its orientation or migration in response to chemotactic stimuli [Malech et al, 1977; Oliver, 1978; Anderson et al, 1982; Schliwa et al, 1982]. The functional role of microtubules in PMN chemotaxis and capping of membrane receptors has been of considerable interest to a number of investigators [Malech et al, 1977; Oliver, 1978; Albertini et al, 1977; Bandman et al, 1974; Gallin and Rosenthal, 1974; Hoffstein et al, 1977]. Some reports have indicated that PMN chemotaxis, but not random migration, is diminished by agents known to depolymerize or prevent assembly of microtubules [Malech et al, 1977; Bandmann et al, 1974; Caner, 1965; Chang, 1975; Valerius, 1978].

Tubulin, the protein dimer which is the major component of microtubules, is subject to a reversible posttranslational modification whereby a tyrosine residue is added to the carboxy-terminal glutamate of tubulin α-chain [Arce et al, 1978; Raybin and Flavin, 1977a]. The enzyme that catalyzes this ATP-dependent reaction, tubulin tyrosine ligase (ligase), has been detected in both vertebrate [Arce et al, 1978; Raybin and Flavin, 1977b] and invertebrate [Kobayashi and Flavin, 1981] tissues. Another distinctly different enzyme, tubulin carboxypeptidase, which is believed to be the major cellular enzyme responsible for the removal of the carboxy-terminal tyrosine of tubulin, has also been detected and characterized

[Agarana et al, 1980; Kumar and Flavin, 1981]. Although the presence of the tyrosine has not been found to affect the assembly of microtubules in vitro, it has been clearly demonstrated that the 6S dimeric tubulin is the substrate for the ligase [Kobayashi and Flavin, 1977], whereas the detyrosinolating enzyme preferentially acts on intact microtubules [Kumar and Flavin, 1981]. Moreover, we have observed changes in the state of tyrosinolation of tubulin in cells undergoing cytoskeletal rearrangements such as in differentiating neuronal cells [Nath and Flavin, 1979] and in human epithelial carcinoma (HeLa) cells during mitosis [Nath et al, 1978]. Recently it has been reported that tubulin α-chain mRNA from both chick [Valenzuela et al, 1981] and rat [Lemischka et al, 1981] brain has a carboxy-terminal tyrosine codon. This elaborate machinery suggests some vital function for the tubulin carboxy-terminal tyrosine, but as yet there are few clues.

We have recently reported a specific, dose-dependent stimulation of tubulin tyrosinolation in rabbit peritoneal neutrophils [Nath et al, 1981] and also in human peripheral blood neutrophils [Nath et al, 1982], as induced by the synthetic peptide chemoattractant, N-formyl-methionyl-leucyl-phenylalanine (fmet-leu-phe). We also studied and compared tubulin tyrosinolation in PMN obtained from normal donors and those from patients with the Chediak-Higashi syndrome (CHS) and with chronic granulomatous disease (CGD). In the present review we discuss some of our results that provide information regarding the modulation of tubulin tyrosinolation in human PMN.

TUBULIN TYROSINOLATION IN NORMAL AND PATIENT PMN

Specific Stimulation of Normal PMN Tyrosinolation by fmet-leu-phe

Since tubulin tyrosinolation is a posttranslational reaction, de novo protein synthesis is not required, and it can be studied in intact functional cells by measuring incorporation of radiolabeled tyrosine into a trichloro-acetic acid (TCA)-insoluble fraction, in the absence of protein synthesis [Nath and Flavin, 1979; Nath et al, 1981; 1982]. To inhibit protein synthesis, the PMN were incubated for 30 min at 37°C with a mixture of antibiotics (protein synthesis inhibitors), prior to the addition of [^{14}C]-tyrosine to the reaction medium. The details of the experimental conditions have been reported recently [Nath and Flavin, 1979; Nath et al, 1981, 1982]. Radiolabeled tyrosine is specifically incorporated into PMN tubulin under the experimental conditions employed in these studies [Nath et al, 1981, 1982]. Figure 1 shows the dose-dependent stimulation of tubulin tyrosinolation in human peripheral blood PMN. Similar stimulation of tyrosine incorporation has been observed in rabbit perito-

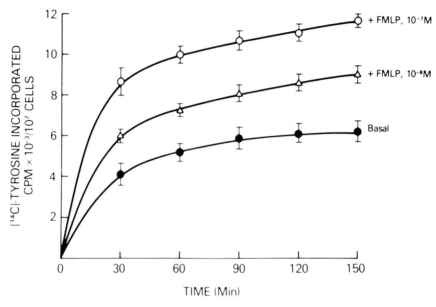

Fig. 1. Stimulation of posttranslational tyrosine incorporation by fmet-leu-phe (FMLP). PMN were preincubated for 30 min at 37°C with protein synthesis inhibitors and FMLP concentrations as indicated. [^{14}C]-tyrosine (5 μCi = 0.01 μmole) was added at 0 min and TCA-insoluble radioactivity was measured at indicated times. The results are means ± SEM of three separate experiments (reproduced from Nath et al [1982]).

neal leukocytes as well [Nath et al, 1981]. The increase in posttranslational incorporation of tyrosine was not due to an increase in protein synthesis in the fmet-leu-phe-stimulated cells nor due to an increased uptake of radioactive tyrosine in these cells [Nath et al, 1981, 1982]. Moreover, the specific activity of ligase, the enzyme which catalyzes the reaction, was not altered and was around 0.01–0.02 nmole min^{-1} mg^{-1} protein, both in resting and stimulated cells [Nath et al, 1982]. This suggested that stimulation of tyrosinolation as induced by the chemoattractant probably involved changes in the functional state of the substrate, ie, PMN tubulin, that resulted in stimulation of tyrosinolation.

Tyrosinolation in PMN From Patients With Chediak-Higashi Syndrome

We have also studied and compared the levels of tubulin tyrosinolation in PMN isolated from CHS patients with those of normal PMN. CHS is a rare disorder with an autosomal recessive pattern of inheritance, characterized by giant lysosomal granules in most granule-containing cells. Clinically, the disease can evolve into a lymphomalike phase with death at an early age; patients not developing lymphoma usually develop severe peripheral neuropathy by age 25 [White and Clawson, 1979]. A major clinical expression is marked susceptibility to pyogenic infections, which, in the absence of a defect in humoral immunity, has directed attention to phagocytic cells and monocytes [White and Clawson, 1979; Wolff et al, 1972]. The patients are neutropenic [Wolff et al, 1972], and PMN and monocytes isolated from CHS patients have been reported to have defective chemotactic responses [Clark and Kimball, 1971; Gallin et al, 1975] and delayed bactericidal activity [Root et al, 1972] despite increased oxidative metabolism [Root and Stossel, 1972]. These abnormalities have often been associated with impaired microtubule function in these cells [Boxer et al, 1977; Oliver and Zurier, 1976]. Comparison of tubulin tyrosinolation in resting PMN obtained from normal donors and from three patients with CHS, demonstrated a two- to threefold higher level of tyrosine incorporation in CHS cells (Fig. 2). The higher tyrosinolation in resting PMN of CHS patients could be further stimulated in the presence of fmet-leu-phe [Nath et al, 1982]. The elevated levels of tyrosinolation in CHS cells was not attributable to a parallel increase in the specific radioactivity of the intracellular pool of tyrosine. Polyacrylamide slab gel electrophoresis of radiolabeled cell samples and subsequent fluorography demonstrated a single radioactive protein band corresponding to authentic [^{14}C]-tyrosinolated tubulin both in normal and CHS cells [Nath et al, 1982]. A quantitative increase in the amount of radioactivity in CHS cell tubulin could also be measured [Nath et al, 1982].

Effect of Ascorbate and Other Reducing Agents on PMN Tubulin Tyrosinolation

Although reports on the efficacy of ascorbate administration in correcting some of the functional abnormalities of PMN related to CHS have been controversial [Boxer et al, 1976; Gallin et al, 1979], there is evidence in the literature to suggest that ascorbate promotes microtubule assembly, both in vivo and in vitro [Boxer et al, 1979], and that it improves defective adherence, chemotaxis, and degranulation of PMN from some CHS patients [Boxer et al, 1976; Goetzel et al, 1974]. Since the reaction which we are studying specifically involves cellular tubulin/microtubules, it was of interest to examine the effect of ascorbate on the posttranslational tyrosine incorporation in PMN of CHS patients. As shown in Figure 3, prior incubation with 10^{-4} M ascorbate normalized the higher resting levels of tyrosine incorporation in CHS cells and, most strikingly, completely inhibited the fmet-leu-phe-induced stimulation both in normal PMN and in PMN obtained from CHS patients. In view of the results shown in Figure 3, we also studied the effect of oral administration of a

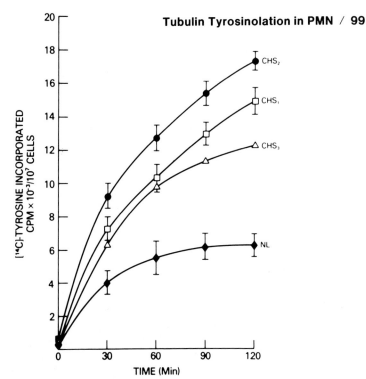

Fig. 2. Comparison of resting levels of tyrosine incorporation in RMN from normal (NL) and three Chediak-Higashi syndrome patients (CHS₁, CHS₂, and CHS₃). Experimental conditions were the same as those described in Figure 1. Cells were preincubated for 30 min with inhibitors of protein synthesis before addition of [^{14}C]-tyrosine. Values are expressed as means ± SEM for three experiments each with CHS₁ and CHS₂, and for six experiments with different normal individuals. The values for CHS₃ are means of duplicate determinations obtained from a single experiment.

large dosage (8 g) of ascorbate to the CHS patients on the rate and extent of posttranslational tyrosine incorporation in their PMN and obtained similar inhibitory results as observed when ascorbate was added in vitro to the reaction medium [Nath et al, 1982].

It is well documented in the literature that changes in oxidative metabolism plays a crucial role in PMN function [Kelbanoff and Clark, 1978a]. Since ascorbate is a powerful reducing agent, we investigated the effect of other reducing agents, such as cysteine, reduced glutathione (GSH), or dithiothreitol (DTT), and found that the various reducing agents also caused inhibition of fmet-leu-phe-induced stimulation as observed with ascorbate (Fig. 4). These results strongly suggested that changes in redox state in stimulated PMN may be linked to the inducton of stimulation of tyrosinolation in these cells.

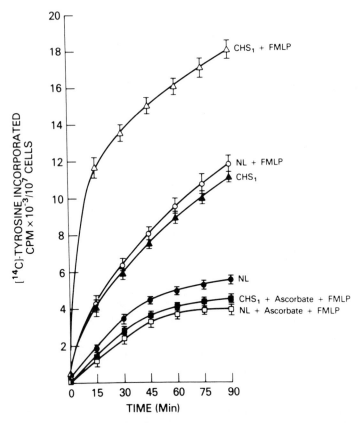

Fig. 3. Effect of addition of ascorbate in vitro on the resting and FMLP-stimulated levels of tyrosine incorporation in normal and CHS_1 PMN. Where indicated, cells were preincubated for 30 min with 10^{-4} M ascorbate and 10^{-7} M FMLP was added at 0 min. Other experimental conditions were the same as in Figure 1. The results are means ± SEM of the three separate experiments.

Effect of Diamide and 2-Cyclohexene-1-one on Tubulin Tyrosinolation

Since GSH peroxidase calatyzed reduction of H_2O_2 plays an important functional role in PMN redox changes [Oliver et al, 1976; Burchill et al, 1978], we studied the effect of diamide, a chemical oxidant of GSH [Kosower et al, 1972], on tubulin tyrosinolation in resting and fmet-leu-phe-stimulated PMN. As reported recently [Nath and Gallin, 1983], 10^{-4} M diamide caused a significant stimulation of both resting and fmet-leu-phe-stimulated incorporation of tyrosine in PMN.

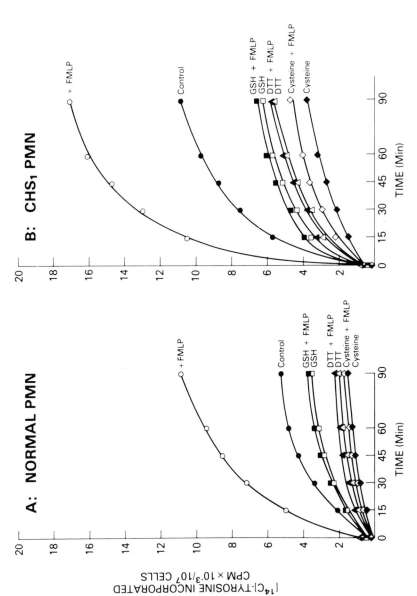

Fig. 4. Effects of various reducing agents on the resting and FMLP-stimulated levels of tyrosine incorporation in normal and CHS$_1$ PMN. Where indicated, PMN were preincubated for 30 min with 10^{-3} M GSH, cysteine, or DTT, and 10^{-7} M FMLP was added at zero min. Other experimental conditions were the same as in Figure 1. A. Normal PMN. B. CHS$_1$ PMN. The results are the means of duplicate determinations (reproduced from Nath et al [1982]).

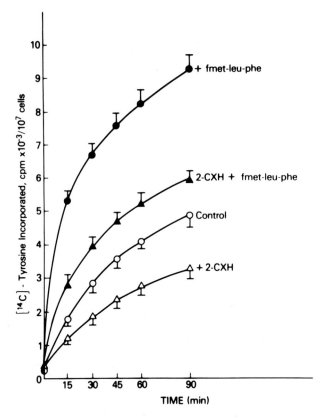

Fig. 5. Effect of 2-cyclohexene-1-one (2-CXH) on resting and fmet-leu-phe-stimulated posttranslational incorporation of tyrosine in normal PMN. Cells at 10^7/ml were incubated for 30 min at 37°C within inhibitors of protein synthesis, in the presence or absence of 10^{-4} M 2-CXH, and [^{14}C]-tyrosine was added at 0 min. Where indicated, 10^{-7} M fmet-leu-phe was also added at 0 min. TCA-insoluble radioactivity was determined at indicated times. Data are means ± SEM of three experiments.

In a recent report, Wedner et al [1981], presented data to indicate that normal physiological GSH levels were necessary for the transducation of the activation signal from the exterior to the interior of the PMN. By using 2-cyclohexene-1-one (2-CXH), which specifically decreases the cellular GSH concentration without increasing oxidized glutathione (GSSG) levels, they could show significant inhibition of PMN function [Wedner et al, 1981]. As shown in Figure 5, prior incubation with 10^{-4} M 2-CXH produced a significant inhibition ($P < 0.02$) of both resting and fmet-leu-phe-stimulated incorporation of tyrosine. The results of our studies with diamide and 2-CXH suggest that GSH homeostasis and thiol-disulfde status of the PMN could also play a regulatory role in the modulation of tubulin tyrosinolation in these cells.

Tubulin Tyrosinolation in PMN From Patients With Chronic Granulomatous Disease

In view of the results obtained with various reducing agents (Fig. 4), we decided to study PMN of patients with CGD, whose oxidative metabolism is severely depressed because of an impaired function of the membrane-associated NADPH-dependent oxidase [Klebanoff and Clark, 1978b]. These cells do not respond to particulate and soluble stimuli with a respiratory burst [Klebanoff and Clark, 1978b; Holmes et al, 1967; Baehner et al, 1976; McPhail et al, 1977] and, therefore, also fail to reduce molecular oxygen to the free radical O_2^-, which results in the generation of other toxic celllar metabolites, such as H_2O_2 and hydroxyl radical [Kelbanoff and Clark, 1978b]. Clinically, these patients have recurrent life-threatening infections with catalase-positive microorganisms. Because of the abnormality in oxidative metabolism, neutrophils from CGD patients should be in a relatively reduced state, in contrast to CHS PMN. Therefore, CGD neutrophils would be expected to have higher concentrations of NADPH and GSH, the effective oxidation of which is coupled to the activation of the oxidative metabolism via the hexose monophosphate shunt (MHPS) pathway [Klebanoff and Clark, 1978b]. Thus, PMN from CGD patients provide an ideal model for study of tubulin tyrosinolation and further investigation of its possible relationship with PMN redox state.

Figure 6 shows the results of our studies with seven CGD patients (four males and three females) and compares them with tyrosinolation levels in resting PMN from normal donors. For purposes of comparison, the figure includes the levels of tyrosinolation observed in resting PMN of CHS patients and shows the results of ten studies with seven different patients with CGD and compares them with that of 30 studies with 15 normal donors and those from 20 studies with three patients with CHS. As compared to normal values, the resting levels of posttranslational incorporation of tyrosine were 35–45% lower ($P < 0.01$) in CGD PMN. The figure also demonstrates the striking difference in the resting levels of tyrosine incorporation between neutrophils from patients with CHS and those with CGD. Figure 7 compares the posttranslational incorporation of tyrosine in normal and CGD PMN in the presence of fmet-leu-phe. In striking contrast to normal and CHS cells, PMN from all seven CGD patients failed to respond to 10^{-7} M fmet-leu-phe, a concentration which induced maximal (two- to threefold) stimulation of tyrosinolation in normal and CHS PMN [Nath et al, 1982].

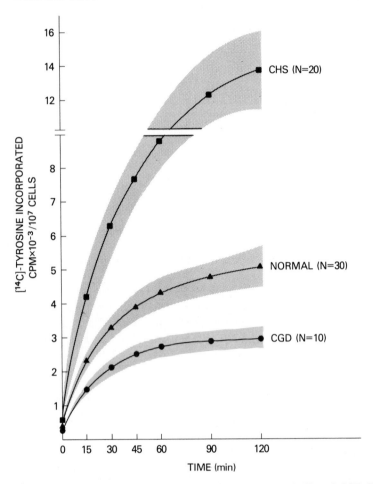

Fig. 6. Comparison of resting levels of tyrosinolation in normal, CHS, and CGD PMN. Cells at 10^7/ml were incubated with inhibitors of protein synthesis for 30 min at 37°C, and [^{14}C]-tyrosine was added at 0 min. TCA-insoluble radioactivity was determined at indicated times. N represents the number of different experiments. The shaded regions represent the range of values.

Effect of Methylene Blue on CGD PMN Tyrosinolation

We have also studied the effect of methylene blue on tubulin tyrosinolation in PMN of CGD patients. Methylene blue, an electron acceptor, oxidizes NADPH to generate $NADP^+$ and, therefore, can correct the oxidative block due to the impaired function of the membrane-associated NADPH-dependent oxidase in PMN of CGD patients [Holmes et al, 1967]. Thus, methylene blue will cause the stimulation of hexosemonophosphate shunt in CGD PMN [Holmes et al, 1967], although it will not

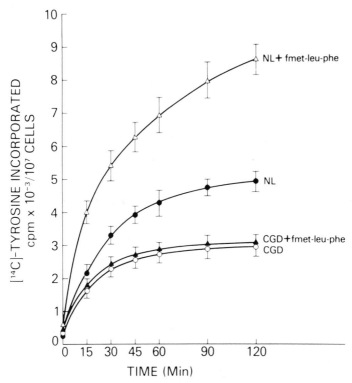

Fig. 7.   Resting and fmet-leu-phe-stimulated posttranslational incorporation of tyrosine in normal and CGD PMN. Cells at 10^7/ml were incubated with inhibitors of protein synthesis for 30 min at 37°C, and [^{14}C]-tyrosine was added at 0 min. Where indicated, 10^{-7} M fmet-leu-phe was also added at 0 min. TCA-insoluble radioactivity was determined at indicated times. Data are means ± SEM of studies in PMN of seven different normal individuals and in seven different patients with CGD (reproduced from Nath and Gallin [1983]).

induce sufficient production of oxygen radicals such as superoxide nor hydrogen peroxide. As recently reported [Nath and Gallin, 1983], addition of 10^{-4} M methylene blue normalized the depressed resting levels of tyrosinolation in PMN of three CGD patients but failed to induce stimulation in the presence of 10^{-7} M fmet-leu-phe.

Effect of Anaerobic Atmosphere and Antioxidants on Tubulin Tyrosinolation in Resting and Stimulated Normal PMN

Anaerobic conditions reportedly inhibit the characteristic respiratory burst in stimulated PMN [Baehner et al, 1976]. In view of the results obtained with PMN from CGD patients, we studied tyrosinolation in normal PMN under anaerobiosis, in an atmosphere of nitrogen. For the

TABLE I. Effect of Anaerobic Atmosphere on the
Posttranslational Incorporation of Tyrosine in PMN

PMN sample[a]	[^{14}C]-tyrosine fixed (cpm/10^7 cells)
In air	
1. Resting	3,680 ± 105
2. +10^{-7} M fmet-leu-phe	8,480 ± 190
In nitrogen	
3. Resting	2,700 ± 85
4. +10^{-7} M fmet-leu-phe	2,500 ± 75

[a]PMN at 10^7/ml were incubated for 60 min with [^{14}C]-tyro-
sine (5 μCi/ml = 0.01 μmole) in the presence of inhibitors of
protein synthesis. As indicated, 10^{-7} M fmet-leu-phe was
added at zero min. TCA-insoluble radioactivity was determined as previously described [Nath and Gallin, 1983]. The
data are means of duplicate determinations (reproduced
from Nath and Gallin [1983]).

experiment shown in Table I, PMN samples ($10'$ cells/ml) were continuously flushed with a gentle stream of nitrogen during the entire incubation
period. The details of the experimental design have been described [Nath
and Gallin, 1983]. As shown in Table I, the fmet-leu-phe-induced
stimulation of tyrosine incorporation, as observed under aerobic conditions, was completely inhibited in PMN under anaerobiosis. We have also
studied the effects of antioxidants such as cysteamine, azide, and 2,3- and
3,5-dihydroxybenzoic acid, on fmet-leu-phe-stimulated tyrosinolation in
PMN and observed significant inhibition of the peptide-induced stimulation [Nath and Gallin, 1983].

These results provide further evidence for a causal relationship between
tubulin tyrosinolation and oxidative-reductive reactions in PMN. Since
both normal PMN under anaerobic conditions and the PMN of CGD
patients fail to elicit the respiratory burst when stimulated [Klebanoff and
Clark, 1978b; Baehner et al, 1976], the present data clearly indicate the
involvement of PMN redox in modulation of tubulin tyrosinolation in
these cells.

CONCLUDING COMMENTS

The studies discussed in the present review clearly indicate a correlation
of tubulin tyrosinolation with the PMN redox state. The functional role of
microtubules in PMN chemotaxis or phagocytosis is not yet completely
defined, but the reported evidence implicate their involvement in cell
orientation and cytoskeletal reorganizations [Malech et al, 1977; Oliver,

1978; Anderson et al, 1982; Schliwa et al, 1982]. More recent morphological studies also suggest that microtubules may provide a "tracking" system for translocation of PMN granules [Ryder et al, 1982; Chandler et al, 1983], which is an essential process in PMN exocytosis [Goldstein et al, 1975; Hoffstein et al, 1977]. Structural interaction of different cytoskeletal components have been demonstrated in a number of cells [Schliwa and Blerkom, 1981], and such interactions could very well play an important functional role in PMN. Although purely speculative at this stage, tyrosinolation of tubulin may play a role in such interactions of the cytoskeletal components and perhaps also in the interaction of microtubules/tubulin with the plasma membrane [Nath and Flavin, 1978, 1980].

PMN oxidative-reductive reactions are coupled to changes in the intracellular free Ca^{2+} concentrations [Becker et al, 1979] and also to the effective operation of the glutathione redox system [Oliver et al, 1976; Burchill et al, 1978]. It is likely, therefore, that the regulation of tubulin tyrosinolation in PMN is quite complex and rests on the delicate equilibrium between a number of physiological and biochemical processes of the cell. Further research is clearly warranted to understand all the complexities of regulation and possible functional role of this intriguing reaction in human PMN.

REFERENCES

Agarana CE, Barra HS, Caputto R (1980): Tubulin-tyrosine carboxypeptidase from chicken brain: Properties and partial purification. J Neurochem 34:114–118.

Albertini DF, Berlin RD, Oliver JM (1977): The mechanism of concanavalin A cap formation in leukocytes, J Cell Sci 26:57–75.

Anderson DC, Wible LJ, Huges BJ, Smith CW, Brinkley BR (1982): Cytoplasmic microtubules in polymorphonuclear leukocytes: Effects of chemotactic stimulation and colchicine. Cell 31:719–729.

Arce CA, Hallak ME, Rodriguez JA, Barra HS, Caputto R (1978): Capability of tubulin and microtubules to incorporate and to release tyrosine and phenylalanine and the effect of the incorporation of these amino acids on tubulin assembly. J Neurochem 31:205–210.

Baehner RL, Boxer LA, Davis J (1976): The biochemical basis of nitroblue tetrazolium reduction in normal human and chronic granulomatous disease polymorphonuclear leukocytes. Blood 48:309–328.

Bandman U, Rydren L, Norberg B (1974): The difference between random movement and chemotaxis. Effects of antitubulins on neutrophil granulocyte locomotion. Exp Cell Res 88:63–73.

Becker EL, Sigman M, Oliver JM (1979): Superoxide production induced in rabbit polymorphonuclear leukocytes by synthetic chemotactic peptides and A23187: The nature of the receptor and the requirement for Ca^{+2}. Am J Pathol 95:81–97.

Boxer LA, Watanabe AM, Rister M, Besch HR, Allen J, Baehner RL (1976): Correction of leukocyte function in Chediak-Higashi syndrome by ascorbate. N Engl J Med 295:1041–1045.

Boxer LA, Rister M, Allen JM, Baehner RL (1977): Improvement of Chediak-Higashi leukocyte function by cyclic guanosine monophosphate. Blood 49:9–17.

Boxer LA, Vanderbilt B, Bousib S, Jersild R, Yang H, Baehner RL (1979): Enhancement of chemotactic response and microtubule assembly in human leukocytes by ascorbic acid. J Cell Physiol 100:119–126.

Burchill BR, Oliver JM, Pearson CB, Leinbach ED, Berlin RD (1978): Microtubule dynamics and glutathione metabolism in phagocytizing human polymorphonuclear leukocytes. J Cell Biol 76:439–447.

Caner JEL (1965): Colchicine inhibition of chemotaxis. Arthritis Rheum 8:757–763.

Chandler DE, Bennett JP, Gomperts B (1983): Freeze-fracture studies of chemotactic peptide-induced exocytosis in neutrophils: Evidence for two patterns of secretory granule fusion. J Ultrastruct Res 82:221–232.

Chang YH (1975): Mechanism of activity of colchicine. II. Effects of colchicine and its analogues on phagocytosis and chemotaxis in vitro. J Pharmacol Exp Ther 194:159–164.

Clark RA, Kimball HR (1971): Defective granulocyte chemotaxis in the Chediak-Higashi syndrome. J Clin Invest 50:2645–2652.

Gallin JI, Rosenthal AS (1974): The regulatory role of divalent cations in human granulocyte chemotaxis. Evidence for an association between calcium exchanges and microtubule assembly. J Cell Biol 62:594–609.

Gallin JI, Klimerman JA, Padgett GA, Wolff SM (1975): Defective mononuclear leukocyte chemotaxis in the Chediak-Higashi syndrome of humans, mink and cattle. Blood 45:863–870.

Gallin JI, Elin RJ, Hubert RT, Fauci AS, Kaliner MA, Wolff SM (1979): Efficacy of ascorbic acid in Chediak-Higashi syndrome (CHS): studies in humans and mice. Blood 53:226–234.

Goetzl EJ, Wasserman SI, Gigli I, Austen KF (1974): Enhancement of random migration and chemotactic response of human leukocytes by ascorbic acid. J Clin Invest 53:813–818.

Goldstein IM, Hoffstein ST, Weissmann G (1975): Mechanisms of lysosomal enzyme release from human polymorphonuclear leukocytes. J Cell Biol 66:647–652.

Hoffstein ST, Goldstein IM, Weissmann G (1977): Role of microtubule assembly in lysosomal enzyme secretion from human polymorphonuclear leukocytes. A reevaluation. J Cell Biol 73:242–256.

Holmes B, Page AR, Good RA (1967): Studies of the metabolic activity of leukocytes from patients with a genetic abnormality of phagocyte function. J Clin Invest 46:1422–1432.

Klebanoff SJ, Clark RA (1978a): The metabolic burst. In Klebanoff SJ, Clark RA (eds): "The Neutorphil: Function and Clinical Disorder." Amsterdam: North-Holland Publishing Company, pp 283–408.

Klebanoff SJ, Clark RA (1978b): Chronic granulomatous disease. In Klebanoff SJ, Clark RA (eds): "The Neutrophil: Function and Clinical Disorders." Amsterdam: North-Holland Publishing Company, pp 641–709.

Kobayashi T, Flavin M (1977): Tubulin-tyrosine ligase purification and application to studies of tubulin structure and assembly. J Cell Biol 75(2, PT. 2):285a.

Kobayashi T, Flavin M (1981): Tubulin tyrosylation in invertebrates. Comp Biochem Physiol [B] 69:387–392.

Kosower EM, Correa W, Kinow BJ, Kosower NS (1972): Glutathione. VII. Differentiation among substrates by the thiol-oxidizing agent, diamide. Biochim Biophys Acta 264:39–44.

Kumar N, Flavin M (1981): Preferential action of a brain detyrosinolating carboxypeptidase on polymerized tubulin. J Biol Chem 256:7678–7686.

Lemischka IR, Farmer S, Racaniello VR, Sharp PA (1981): Nucleotide sequence and evolution of a mammalian α-tubulin messenger RNA. J Mol Biol 151:101–120.

Malech HL, Root RK, Gallin JI (1977): Structural analysis of human neutrophil migration. J Cell Biol 75:666–693.

McPhail LC, DeChatelet LR, Shirley PS, Wilfert C, Johnston RB, McCall CE (1977): Deficiency of NADPH oxidase activity in chronic granulomatous disease. J Pediatr 90:217–217.

Nath J, Flavin M (1978): A structural difference between cytoplasmic and membrane-bound tubulin of brain. FEBS Lett 95:335–338.

Nath J, Flavin M (1979): Tubulin tyrosylation in vivo and changes accompanying differentiation of cultured neuroblastoma-glioma hybrid cells. J Biol Chem 254:1505–1510.

Nath J, Flavin M (1980): An apparent paradox in the occurrence, and in the in vivo turnover, of C-terminal tyrosine in membrane bound tubulin of brain. J Neurochem 35:693–706.

Nath J, Gallin JI (1983): Studies in normal and chronic granulomatous disease neutrophils indicate a correlation of tubulin tyrosinolation with the cellular redox state. J Clin Invest 71:1273–1281.

Nath J, Flavin M, Schiffmann E (1981): Stimulation of tubulin tyrosinolation in rabbit leukocytes evoked by the chemoattractant formyl-methionyl-leucyl-phenylalanine. J Cell Biol 91:232–239.

Nath J, Flavin M, Gallin JI (1982): Tubulin tyrosinolation in human polymorphonuclear leukocytes: Studies in normal subjects and in patients with the Chediak-Higashi sundrome. J Cell Biol 95:519–526.

Nath J, Whitlock J, Flavin M (1978): Tyrosylation of tubulin in synchronized Hela cells. J Cell Biol 79:294a.

Oliver JM (1978): Cell Biology of leukocyte abnormalities-membrane and cytoskeletal function in normal and defective cells. Am J Pathol 93:221–278.

Oliver JM Zurier RB (1976): Correction of characteristic abnormalities of microtubule function and granule morphology in Chediak-Higashi syndrome with cholinergic agonists: Studies in vitro in man and in vivo in the beige mouse. J Clin Invest 57:1239–1247.

Oliver JM, Albertini DF, Berlin RD (1976): Effects of glutathione-oxidizing agents on microtubule assembly and microtubule-dependent surface properties of human neutrophils. J Cell Biol 71:921–932.

Raybin D, Flavin M (1977a): Enzyme which specifically adds tyrosine to the α-chain of tubulin. Biochemistry 16:2189–2194.

Raybin D, Flavin M (1977b): Modification of tubulin by tyrosylation in cells and extracts and its effect on assembly in vitro. J Cell Biol 73:492–504.

Root RK, Stossel T (1972): Functional comparison of Chediak-Higashi syndrome and chronic granulomatous disease leukocytes. Blood. Birth Defects: Original Article Series Vol VIII, No. 3.

Root RK, Rosenthal AS, Balestra DJ (1972): Abnormal bactericidal, metabolic and lysosomal functions of Chediak-Higashi syndrome leukocyte. J Clin Invest 51:649–665.

Ryder MI, Niederman R, Taggart EJ (1982): The cytoskeleton of human polymorphonuclear leukocytes: Phagocytosis and degranulation. Anat Rec 203:317–327.

Schliwa M, Blerkom JV (1981): Structural interaction of cytoskeletal components. J Cell Biol 90:222–235.

Schliwa M, Pryzwansky KB, Euteneuer U (1982): Centrosome splitting in neutrophils: An unusual phenomenon related to cell activation and motility. Cell 31:705–717.

Valenzuela P, Quiroga M, Zaldiver J, Rutter WJ, Kirschner MW, Cleveland DW (1981):

Nucleotide and corresponding amino acid sequences encoded by α and β tubulin mRNAs. Nature 289:650–655.

Valerius NH (1978): In vitro effects of colchicine on neutrophil granulocyte locomotion. Acta Pathol Microbiol Scand 86:149–154.

Wedner HJ, Simchowitz L, Stenson WF, Fischman CM (1981): Inhibition of human polymorphonuclear leukocyte function by 2-cyclohexene-1-one. A role of glutathione in cell activation. J Clin Invest 68:535–543.

White JG, Clawson CC (1979): The Chediak-Higashi syndrome: Microtubules in monocytes and lymphocytes. Am J Hematol 7:349–356.

Wolff SM, Dale DC, Clark RA, Root RK, Kimball H (1972): The Chediak-Higashi sundrome: Studies of host defense. Am Intern Med 76:293–306.

White Cell Mechanics: Basic Science and
Clinical Aspects, pages 111–127

Some Functional Effects of Physical Changes in the Membranes of White Cells

M. L. Karnovsky, J. A. Badwey, J. T. Curnutte, J. M. Robinson, C. B. Berde, and M. J. Karnovsky

Departments of Biological Chemistry (M.L.K., J.A.B.), Pediatrics (J.T.C., C.B.B.), and Pathology (J.M.R., M.J.K.), Harvard University Medical School, Boston, Massachusetts 02115

INTRODUCTION

Membranes play a cardinal role in the functions of phagocytic cells. This rather obvious statement becomes even more evident when one contemplates the phagocytic process per se. The mechanical phenomena of pseudopod formation, envelopment of the object, internalization of the vesicle (phagosome), and fusion of lysosomal granules with the phagosome to release enzymes and form the phagolysosomes are all clearly functions that involve physical changes in the membrane. We may call these "macromembrane effects," in our present context. Phenomena that precede these more mechanical membrane functions, the subtle matters involving binding sites, receptors, and other membrane structures (usually protein or glycoprotein) that are involved in chemotaxis, particle recognition, binding of particles to the cell surface, etc, have been known and studied for many years. Lectin-binding sites, ectoenzymes, and the receptors for some of the substances of particular significance in the context of this chapter (eg, phorbol esters) are of more recent interest and exploitation [Klebanoff and Clark, 1978; Cohen et al, 1980].

The topic of this chapter is the role of membranes of phagocytic leukocytes in the "respiratory burst" that often accompanies the stimulation of the cells in various ways [Babior and Crowley, 1983; Badwey and Karnovsky, 1980]. This burst may serve as the indicator of functional change in the membrane. It will emerge later that several lines of evidence, and several models, point to the involvement of physical phenomena in the membrane itself that facilitate enzymatic function of membrane components in promoting the burst. These we may call "micromembrane effects."

The "respiratory burst" involves increased oxygen uptake, and increased production of oxygen radicals, or activated species, such as superoxide (O_2^-), hydrogen peroxide (H_2O_2), hydroxyl radical ($OH\cdot$), and perhaps singlet oxygen ($^1O_2$). The reactions involved in the reduction of O_2, and the interactions of the active species themselves, have been extensively reviewed recently [Babior et al, 1973; Babior and Crowley, 1983; Badwey and Karnovsky, 1980].

The agents that induce the augmented production of active oxygen species by phagocytic leukocytes are quite diverse. They may be, as mentioned, phagocytizable particles such as polystyrene latex, bacteria, or fragments of yeast cell wall (zymosan). Such particles may require opsonization, that is, the attachment of certain serum proteins, or they may not, as in the case of latex particles. Contact of the phagocytic leukocytes with rather large objects, such as other mammalian cells (eg, tumor cells) or parasites (eg, schistosomes) that are targets for their activities, or with solid surfaces, may also have the effect of stimulating the oxidative metabolism of the former cells (reviewed in Badwey and Karnovsky [1980], and in Badwey et al [1983]). All those mentioned may be referred to as "insoluble stimuli," and the degree to which they are effective in the context of inducing increased oxygen uptake and oxygen radical formation varies.

Among those we may classify as "soluble stimuli" are some that are not truly soluble, but may be finely dispersed. Such membrane perturbants such as digitonin or deoxycholate may be detergents or surfactants. Phorbol esters, particularly phorbol-12-myristate-13-acetate (PMA); lectins such as concanavalin A; the calcium ionophore A23187; chemotactic peptides (eg, N-formyl-L-methionyl-L-leucyl-L-phenylalanine); and the fluoride anion have also been widely used. In the case of soluble stimulating agents, cytochalasins are often used in conjunction with the stimulus itself, and this matter will be referred to later [Klebanoff and Clark, 1978; Badwey et al, 1983a]. Soluble agents that have attracted attention recently are the cis-polyenoic fatty acids, such as arachidonic acid, and the discussion of the effect of these comprises a main component of this communication [eg, Badwey et al, 1981; Kakinuma and Minakami, 1978].

The focus of this chapter will be largely on granulocytes, which have been the most widely used cellular models (See below). In the case of macrophages the history of the cell population isolated dictates whether or not a respiratory burst is observed upon stimulation. For example, resident macrophages of the mouse peritoneum do not show a respiratory burst upon stimulation, whereas those cells elicited with some inflammatory agents (eg, caseinate, proteose peptone) or obtained after infection of the animal with an organism such as *Listeria monocytogenes,* or Calmette-Guérin bacillus (BCG), are active in this regard [Nathan and Root, 1977; Johnston et al, 1978; Badwey et al, 1983b]. Species differences, together with questions of locale of harvesting and the history of the animal, render it necessary to make comparisons between observations very judiciously, especially when variation in the nature of the stimulus and the conditions of the experiment itself occur. On the other hand, it is clear that there is a large body of information that indicates generality in the nature of the respiratory burst, at the cellular membrane and molecular level [Badwey and Karnovsky, 1980].

ASPECTS OF THE RESPIRATORY BURST

Though the phenomenon has now been known for close to half a century and has been worked on very actively, particularly during the past two decades, the exact mechanism of the respiratory burst is currently incompletely understood. There are differences between guinea pig and human granulocytes that have sometimes obscured the issues. As mentioned above, there are also differences between granulocytes and macrophages. The body of information available for the latter cells is small, compared with that for the former. Thus, the granulocyte has been used as the principal model for elucidating the enzymatic basis of the phenomenon. Presumably this is due to the ease of harvesting large numbers of granulocytes and the vigor of the response of those cells, particularly from humans.

In the case of the human granulocyte, an "NAD(P)H oxidase" is commonly invoked as the principal enzyme in the burst (for review see Rossi et al [1980]). Exactly what the pathway of electrons may be has not been determined. A cytochrome of the b series described by British workers as "cytochrome b_{-245}" together with an appropriate reductase that carries the electrons from the reduced pyridine nucleotide to the cytochrome is apparently involved [Segal et al, 1983; M.L. Karnovsky, 1983]. A recently described cytochrome b_5 reductase is a potential candidate for this role [Badwey et al, 1983c], but linkage to oxygen is not definitively established. A mechanism by which the reduced cytochrome is oxidized, or autooxidized, with the formation of O_2^- and H_2O_2 must still be defined. As will be seen below, there is strong evidence that the process described occurs in the membrane of the cell. In fact, the NAD(P)H

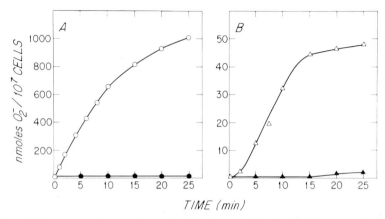

Fig.1. Time course of superoxide production of neutrophils. Each of the curves depicted is representative of the time course of superoxide production by phagocytizing human blood (O, part A) and elicited guinea pig (Δ, part B) neutrophils. The corresponding closed symbols (●, ▲) are for the resting (nonphagocytizing) cells. The concentrations of cells were 5×10^5 cells/ml for human neutrophils, and 2×10^7 cells/ml for elicited guinea pig neutrophils. (Note: O_2^- production was linear over a range of cell concentrations of 2.5×10^5 to 2×10^7 cells/ml.) Opsonized zymosan served as the activating agent at a concentration of 3 mg/ml.

oxidase system is found in the membrane fragments upon fractionation of the cells [Babior and Crowley, 1983].

Stimulation of Oxygen Uptake and Radical Formation During Phagocytosis

Figure 1 illustrates the respiratory burst, measured as superoxide production, by two types of granulocytes, ie, from human blood or guinea pig peritoneum. It is clear that the former cells are far more active in mounting such a burst than are the latter cells (about 20-fold) when stimulated with opsonized zymosan. Furthermore, it is obvious that the resting, unstimulated, cells produce little superoxide [Badwey et al, 1980]. The question arises as to where this superoxide, and, consequently, peroxide, are formed and released. Unfortunately no cytochemical technique has yet been devised for the determination of the cellular location of superoxide release, but such a method has been worked out for peroxide [Briggs et al, 1977]. It involves the conversion of cerous ion to cerium perhydroxide in the presence of hydrogen peroxide [Briggs et al, 1975]. Cerium perhydroxide is extremely electron dense and insoluble and, therefore, has suitable attributes for a cytochemical marker [M.J. Karnovsky et al, 1981].

Figure 2 illustrates the cellular locales of peroxide release upon stimulation of human granulocytes with zymosan [Badwey et al, 1980]. During phagocytosis it is clear that the reaction product is deposited on the plasmalemma and on the membrane lining the phagosome formed by invagination of the plasmalemma. Thus, one may regard this as the basic cytochemical evidence for the involvement of the membrane structure of the cells in the respiratory function, by indicating the sites of release of active products.

In Figure 2 one may also note the effect of stimulation of the same cells with a nonparticulate agent, phorbol myristate acetate (PMA). It is clear that, once more, there is involvement of the plasmalemma, and in the absence of phagosomes (and phagolysosomes), there is clear involvement of vesicles that have been formed and internalized. Deposition of reaction product in these vesicles may be noted. The picture for the guinea pig or mouse cell is very similar.

With respect to macrophages stimulated with PMA it should be remarked that deposition of reaction product was rarely observed in the case of resident peritoneal macrophages and was restricted to internalized vesicles when it occurred. Elicited macrophages exhibited many more peroxide-positive cells and the deposition was again limited to interior vesicles induced by PMA. In the case of the immunologically activated cells from animals infected with *L. monocytogenes* a small subpopulation was noted in which the cells were rather large and deposition occurred both on interior vesicles and on the plasmalemmal surface [Badwey et al, 1983b].

SOLUBLE MEMBRANE PERTURBANTS OTHER THAN PMA

In the studies above it is clear that phagocytosis and stimulation by PMA both involve the formation of plasmalemmal vesicles of different sizes. Digitonin and deoxycholate also cause perturbation of the membrane and the respiratory burst [Graham et al., 1967; Cohen and Chovaniec, 1978]. The case of digitonin is of particular interest. It, too, stimulated the cyanide-insensitive respiratory burst in guinea pig granulocytes and the grossly increased incorporation of inorganic ^{32}P into acidic glycerophosphatides—an effect previously observed during phagocytosis itself (ie, ingestion of solid particles) [Karnovsky and Wallach, 1961]. Although the respiratory burst generally accompanies inward vesiculation, digitonin caused the formation of *protuberances* of the cell membrane as shown in Figure 3, rather than the generation of vesicles. Thus, both inward and outward "macromembrane effects" are accompanied by the respiratory burst. Similar observations apply to deoxycholate as a

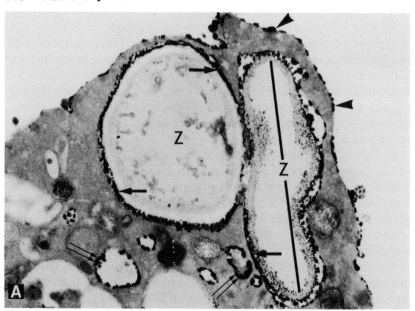

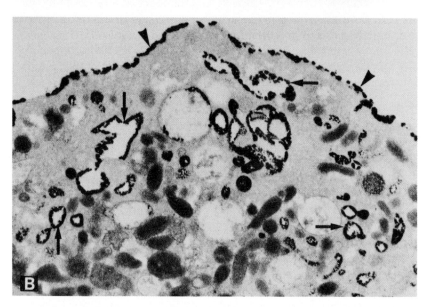

Fig. 2. A. Electron micrograph of a portion of a human neutrophil that phagocytized opsonized zymosan particles (Z) prior to incubation in the cytochemical medium for H_2O_2 localization. The electron-dense reaction product is localized to the phagosome membrane (bold arrows) and the cell surface (arrowheads). Reaction product can also be detected in vesicular structures (thin arrows), which are observed only following phagocytosis. ×27,000. B. Electron micrograph of a portion of human blood neutrophil that was stimulated with PMA prior to incubation in the cytochemical medium for H_2O_2 localization. Reaction product is evident on the cell surface (arrowheads) and within small vesicular structures (arrows). ×26,000.

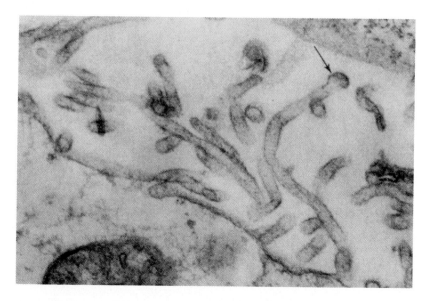

Fig. 3. Cylindrical projections from a mononuclear cell incubated with digitonin. One projection appears to have a bulbous end (arrow). ×87,000.

perturbant and to macrophages from guinea pigs. It may be noted that inhibitors of the respiratory burst did not interfere with the morphological changes described.

EFFECTS OF FATTY ACIDS ON THE RESPIRATORY PATTERN OF LEUKOCYTES

The aliphatic monocarboxylic acids have provided an extraordinarily useful tool in probing the nature of the sequence of events that underlie the respiratory burst at the level of the cell membrane. This is particularly true of the cis-polyunsaturated fatty acids, especially arachidonate [Badwey et al, 1981]. As may be seen in Figure 4, arachidonate is a stimulus for the respiratory burst in human granulocytes comparable to phorbol myristate acetate and opsonized zymosan in effectiveness. These are the most potent stimuli previously used. Second, the effect is a function of the number of double bonds and their stereochemistry. The trans geometric isomers and saturated fatty acids are inactive.

In the context of this paper, a major matter is whether a fatty acid that causes an increased respiratory burst also causes the deformation or perturbation of the membrane(s) of the cell—ie, a "macro membrane

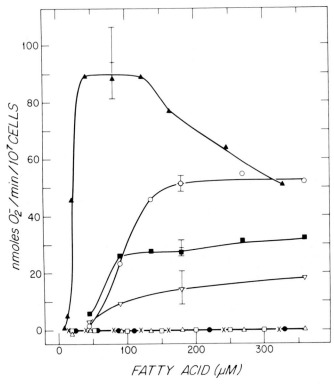

Fig. 4. Ability of various fatty acids to stimulate superoxide production by human neutrophils. Superoxide production was monitored after stimulation with different concentrations of fatty acids. The symbols correspond to the following fatty acids: arachidonate (▲), γ-linolenate (O), linoleate (■), oleate (∇), myristate (●), palmitate (□), linolelaidate (△), and 11-eicosaenoate acid (X). Each curve depicted is representative of the dose-response relationship observed for 1.5×10^6 cells/ml.

effect." Figure 5 shows a comparison between human granulocytes that are unstimulated and those that have been treated with arachidonate to induce the respiratory burst. It is obvious that in the latter case, there has been a considerable perturbation, in a gross physical sense (macro) of the plasmalemma, particularly. It would also be important to know whether the plasmalemma is perturbed in the same way by such a small soluble stimulating agent as fluoride ion [Curnutte et al, 1979].

A most useful aspect of the respiratory phenomenon involving arachidonate is its ready reversibility. Addition of delipidated albumin to a suspension of human granulocytes stimulated with arachidonate returns the cell respiration to unstimulated levels in a stoichiometric fashion, as shown in Figure 6. Further addition of arachidonate and of albumin alternately stimulate and reverse the phenomenon. This reversibility confirms what was earlier noted with fluoride as a stimulus, but in this case involves much less complex operations than were formerly necessary [Curnutte et al, 1979]. Reversibility of the respiratory burst has important implications for understanding the mechanism of O_2^- production (see below). One matter that emerges from these experiments is that the polyenoic fatty acid itself, and not one of the possible products of its metabolism (eg, prostaglandins, hydroxy fatty acids, phospholipids, triglycerides), apparently can initiate and maintain the respiratory phenomenon.

"Microeffects" of Fatty Acids on the Plasma Membrane

A good deal of attention has been paid in recent years to the possibility of the existence of lipid domains in the plasmalemma, ie, in the lipid bilayer of the membrane [M.J. Karnovsky et al, 1982]. It has also been specified that saturated or cis unsaturated fatty acids may preferentially partition between different domains of the lipid bilayer with differing effects on the physical nature of the membrane, such as apparent fluidity, ordering of the head groups, etc. Trans unsaturated fatty acids are reported to behave in the main like saturated fatty acids of equal chain length. Shorter chain fatty acids may have some effects more akin to those of unsaturated fatty acids. It was of considerable importance in the current investigation to determine what the physical effects of polyenoic fatty acids might be when they are inserted into the membrane of phagocytic leukocytes. Second, it would be useful to compare the pattern of effects by various fatty acids in this context with the patterns that have been noted in the case of such biological phenomena as lymphocyte capping, platelet aggregation, etc (for review see M.J. Karnovsky [1983]). Finally, actual monitoring of the physical changes (eg, fluidity) in the membrane induced by fatty acids would be important. In order to examine these effects, we employed as fluorescent probes two naturally occurring 18 carbon conjugated polyenoic fatty acids, "cis" parinaric acid (9, 11, 13, 15 cis-trans-trans-cis octadecatetraenoic acid) and "trans" parinaric acid (9, 11, 13, 15 all-trans octadecatetraenoic acid). The studies of Sklar and co-workers [eg, Sklar et al, 1979] in model systems and in cells have established these compounds as useful nonperturbing probes of bilayer structure. Trans parinaric acid partitions preferentially into solid domains ($K_p^{s/f} \doteq 4$), whereas cis parinaric acid partitions with a slight preference for fluid domains ($K_p^{s/f} \doteq 0.7$) [Sklar et al, 1979]. In model systems with mixtures of phospholipids, the polarization of parinaric acid fluorescence

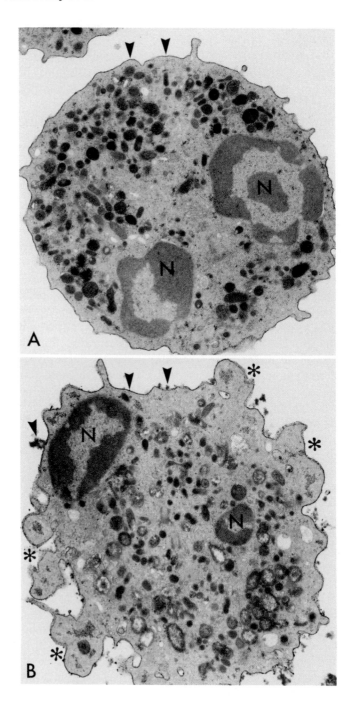

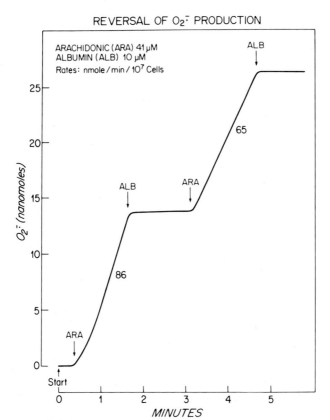

REVERSAL OF O_2^- PRODUCTION

ARACHIDONIC (ARA) 41 μM
ALBUMIN (ALB) 10 μM
Rates: nmole / min / 10^7 Cells

O_2^- (nanomoles)

MINUTES

Fig. 6. Reversal of arachidonic acid-stimulated O_2^- production by delipidated albumin. Human granulocytes were stimulated with arachidonic acid (ARA) at the points indicated by arrows. After approximately 1 minute, delipidated albumin (ALB) was added, as indicated (arrow). The amounts involved are listed on the chart. It will be noted that the albumin switched off the effect immediately and that the effect could be reinstituted by additional arachidonate.

Fig. 5. Effects of treatment of human neutrophils with arachidonic acid under conditions that stimulate the respiratory burst. A. Control incubated in cytochrome c-containing medium. The cytochrome c present at the cell surface (arrowheads) presumably results from cross-linking of this protein to the cell during glutaraldehyde fixation. Note that the control cell is rounded with occasional microvilli. B. Arachidonic acid-treated cell. Cell was in cytochrome c-containing medium during treatment, as in A. Note that this cell is not round, but has numerous blunt projections (*) which often appear to be pinching off from the cell body. Granules are swollen and fused, and glycogen is concentrated in the lobes of cytoplasm. Another feature which was often seen, and is evident in the lower left-hand corner of B, is that many blebs also contain granules.

has been shown to be a useful means for determining phase diagrams, and increases or decreases in fluorescence polarization in two-phase regions are associated with increases or decreases in the fraction of lipid which is solid. Thus, to the extent that biological membranes behave as coexisting patches of solid and fluid lipid, changes in the parinaric acid fluorescence polarization ratio should be correlated with changes in the fraction of the membrane which is fluid or solid. These probes are particularly appropriate in the current situation because they do not activate superoxide production or oxygen consumption and because, as in model systems, cis and trans parinaric acid can be expected to partition into environments similar to those of group A and group B fatty acids [M.J. Karnovsky et al, 1982, 1983], respectively.

We have shown that arachidonic and linoleic acids, which stimulate superoxide production, decrease the polarization of cis and trans parinaric acid fluorescence, while myristic acid, which does not activate superoxide production, increases the polarization of cis and trans parinaric acid fluorescence. Control experiments ruled out interference due to micelle formation or changes in probe binding. These apparent changes in the fraction of membrane which is fluid occurred at concentrations of arachidonate and linoleate well below those required for maximal activation of superoxide production. Changes in the fluidity of the membrane might permit membrane proteins that are components of the system responsible for the respiratory burst to make contact with each other, and thus activate the enzyme complex. A model for this "triggering phenomenon" has been under consideration in many laboratories and by many authors. In view of our own results, the outline of such a model will be presented later in this paper.

THE ROLE OF THE CYTOSKELETON IN CONJUNCTION WITH THE PLASMALEMMA

The effect of cytochalasin E on the respiratory burst induced by PMA has been studied [Badwey et al, 1982]. This cytochalasin provides a useful tool in attempting to define the possible partnership between the cytoskeleton and the plasmalemma in this phenomenon, because it has been reported to have a particular selectivity for the dissolution of the cytoskeleton [eg, Yahara et al, 1982]. The effects of many soluble stimuli have previously been studied in the presence of cytochalasin B or E. This was to provide a means of avoiding the internalization of products of the respiratory burst by blocking vesicle formation, thus allowing such products to be detectable and measureable in the external medium after release from the plasmalemma. Several of the stimuli that have been used result in a lag period before the increase in the rate of respiration (or superoxide production) is evident. This lag period may vary from several seconds in the case of arachidonate to several minutes with fluoride [Curnutte et al, 1979]. One would assume that each stimulus may *operate at a specific point* in the sequence of steps that results in the respiratory burst (see model below). It is possible that the lag period represents the period involved in the establishment of new relationships between the plasmalemma and the cytoskeleton. For example, it is thought that the latter essentially forms a barrier to the fusion of granules with the plasmalemma membrane. Rapid dissolution of cytoskeleton with the aid, for example, of cytochalasin E might facilitate the collaboration of the two subcellular entities.

Figure 7 illustrates that in the presence of cytochalasin E, the lag phase normally observed in the respiratory burst when the stimulus is PMA, is

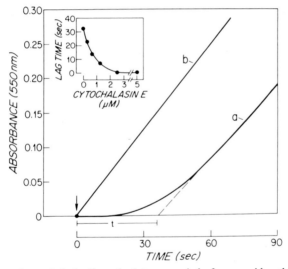

Fig. 7.   Effect of cytochalasin E on the latency period of superoxide release by human neutrophils. Reaction progress curves demonstrate the effect of cytochalasin E on the superoxide-dependent reduction of cytochrome c by human neutrophils stimulated with PMA (1 μg/ml), which was added last to initiate the reactions (bold arrows). Curve a shows the initial portion of the progress curve for 1.5×10^6 cells/ml in the absence of cytochalasin E. The time (t) required to reach the maximal activity (lag time) was approximated by back-extrapolation to zero absorbance change. Incubation of cells with cytochalasin E (2.5 μM) prior to stimulation abolished the lag (curve b). The inset shows the effect of various concentrations of cytochalasin E on the lag time. Cells were incubated with cytochalasin E for 10 minutes prior to stimulation.

totally eliminated. Furthermore, electron micrographs indicate that the cytoskeleton disappears under these conditions, ie, in the presence of cytochalasin E.

A MODEL FOR THE MACHINERY OF THE RESPIRATORY BURST

The following ideas constitute a tentative model of the system that underlies the respiratory burst. It is not totally original but incorporates ideas of others as well as our own. Further, the explanations and points made depend heavily on effects of potential inhibitors as these affect stimulation of the cells by each of a variety of agents. We believe each of these stimulating agents may set the triggering chain in action by exerting its influence at different points along that chain. For example, arachidonic acid may act distally, and opsonized zymosan proximally, to the beginning of that chain. In other terms, in a sequence

$$
\begin{array}{ccc}
\text{I} & \text{II} & \text{NAD(P)H} + \text{O}_2 \\
\text{trigger system} & \text{oxidase complex} & \\
[\text{a} \rightarrow \text{b} \rightarrow \text{c} \rightarrow \text{d} \rightarrow \rightarrow] \xrightarrow[\text{activates}]{} & [\text{components (1 to n)}] & \rightarrow \text{O}_2^-
\end{array}
$$

opsonized zymosan may exert its effect at "a" and a cis-polyenoic fatty acid at "d." Further, each of various inhibitors may stop the burst by interfering with, or by blocking, one step or another (eg, "a" → "b", or "b" → "c", etc.).

Some of these components of the trigger system (I) may be anchored by linkage to the cytoskeleton which may dictate their mobility. The cytoskeleton may also be manipulated by chemical or biological additives (eg, cytochalasin E).

The enzyme system that actually transfers electrons to oxygen (II) to form O_2^- (and secondarily H_2O_2, OH·, and $^1O_2$) utilizes a reduced pyridine nucleotide as the source of those electrons—replenished by metabolic reactions. Evidence exists that this enzyme system consists minimally of a b-cytochrome and, an NAD(P)H — cytochrome b reductase [Segal et al, 1983; Badwey et al, 1983c]. The reduced form of the cytochrome may require a terminal enzymatic step to reduce O_2 or may autooxidize, as indicated earlier, with the formation of O_2^-. A quinone may also participate [Cranford and Schneider, 1982]. (Substances that may block the enzymatic chain (II) per se are not considered here).

The "trigger system" (I) may be considered as facilitating contact of entities of the oxidase system (II) (eg, the reductase and the cytochrome).

This can be by fluidization of the lipid bilayer (eg, artificially by added arachidonate or naturally by enzymatic release of another lipid substance with fluidizing capability). To add a degree of complexity, it has been clear for years that Ca^{+2} plays a crucial role—and we have recently obtained evidence that calmodulin is involved in regulating calcium's part in the sequence—eg, possibly at the level of activation of a phospholipase that produces the lipid(s) involved in the trigger mechanism, as mentioned earlier. The reversibility of the arachidonic acid induced burst is viewed as contraindicating involvement of a protease in the general phenomenon, a matter suggested by others [eg, Kitagawa et al, 1980].

Clearly, our thinking is directed largely at the "micromembrane effects" as instigators of the respiratory burst. Of the "macromembrane effects" we can say little except that they too may depend upon the "microeffects." Their development as a *basis* for the burst is problematical—they are probably secondary phenomena. However, they must involve the cytoskeleton. Ultimately, completion of the entire jigsaw puzzle of the respiratory burst of phagocytic leukocytes will depend upon fitting together data on receptors, membrane structure, cytoskeleton elements, substances that regulate calcium ion concentrations and movements, and the appropriate enzymes. Progress has been slow, but the times are currently propitious for rapid development of understanding in this field.

ACKNOWLEDGMENTS

Studies cited from our laboratories were supported by U.S. Public Health grants, NIH Nos. AI-03260 and AI-17945.

Figures 1 and 2 are reprinted with permission of the Journal of Cellular Physiology; Figure 3 is reprinted with the permission of the Journal of Cell Biology; Figure 4 with the permission of the Journal of Biological Chemistry; and Figure 7 with that of Biochemical and Biophysical Research Communications.

REFERENCES

Babior BM, Kipnes RS, Curnutte JT (1973): Biological defense mechanisms. The production by leukocytes of superoxide, a potential bactericidal agent. J Clin Invest 52:741–744.
Babior BM, Crowley CA (1983): Chronic granulomatous disease and other disorders of oxidative killing by phagocytes. In Stanbury JB, Wyngaarden JB, Fredrickson DS, Goldstein JL, Brown MS (eds): "The Metabolic Basis of Inherited Disease." New York: McGraw-Hill, pp 1956–1985.

Badwey JA, Karnovsky ML (1980): Active oxygen species and functions of phagocytic leukocytes. Annu Rev Biochem 49:695–726.

Badwey JA, Curnutte JT, Robinson JM, Lazdins JK, Briggs RT, Karnovsky MJ, Karnovsky ML (1980): Comparative aspects of the oxidative metabolism of neutrophils from human blood and guinea pig peritonea: Magnitude of the respiratory burst, dependence upon stimulating agents, and localization of the oxidases. J Cell Physiol 105:541.

Badwey JA, Curnutte JT, Karnovsky ML (1981): Cis-polyunsaturated fatty acids induce high levels of superoxide production by human neutrophils. J Biol Chem 256:12640–12643.

Badwey JA, Curnutte JT, Berde CB, Karnovsky ML (1982): Cytochalasin E diminishes the lag phase in the release of superoxide by human neutrophils. Biochem Biophys Res Commun 106:170–174.

Badwey JA, Robinson JM, Karnovsky MJ, Karnovsky ML (1983a): Reduction and excitation of oxygen by phagocytic leukocytes: Biochemical and cytochemical techniques. In Weir DM, Herzenberg LA, Blackwell CC, Herzenberg LA (eds): "Handbook of Experimental Immunology," Vol II. Edinburgh: Scientific Publications Ltd (in press).

Badwey JA, Robinson JM, Lazdins JK, Briggs RT, Karnovsky MJ, Karnovsky ML (1983b): Comparative biochemical and cytochemical studies on superoxide and peroxide in mouse macrophages. J Cell Physiol 115:208–216.

Badwey JA, Tauber AI, Karnovsky ML (1983c): Properties of NADH-cytochrome b_5 reductase from human neutrophils. Blood (in press).

Briggs RT, Drath DB, Karnovsky ML, Karnovsky MJ (1975): Localization of NADH oxidase on the surface of human polymorphonuclear leukocytes by a new cytochemical method. J Cell Biol 67:566–586.

Briggs RT, Karnovsky ML, Karnovsky MJ (1977): Hydrogen peroxide production in chronic granulomatous disease. A cytochemical study of reduced pyridine nucleotide oxidases. J Clin Invest 59:1088.

Cohen HJ, Chovaniec ME (1978): Superoxide generation by digitonin-stimulated guinea pig granulocytes: A basis for a continuous assay for monitoring superoxide production and for the study of the activation of the generating system. J Clin Invest 61:1081.

Cohen MS, Metcalf JA, Root RK (1980): Regulation of oxygen metabolism in human granulocytes: Relationship between stimulus binding and oxidative response using plant lectins as probes. Blood 55:1003.

Cranford DR, Schneider DL (1982): Identification of ubiquinone-50 in human neutrophils and its role in microbicidal events. J Biol Chem 257:6662–6668.

Curnutte JT, Babior BM, Karnovsky ML (1979): Fluoride-mediated activation of the respiratory burst in human neutrophils. A reversible process. J Clin Invest 63:637.

Goldstein IM, Cerqueira M, Lind S, Kaplan HB (1977): Evidence that the superoxide-generating system of human leukocytes is associated with the cell surface. J Clin Invest 59:249–254.

Graham RC, Karnovsky MJ, Shafer AW, Glass EA, Karnovsky ML (1967): Metabolic and morphological observations on the effect of surface-active agents on leukocytes. J Cell Biol 32:629.

Johnston RB, Godzik CA, Cohn ZA (1978): Increased superoxide anion production by immunologically activated and chemically elicited macrophages. J Exp Med 148:115–127.

Kakinuma K, Minakami S (1978): Effects of fatty acids on superoxide radical generation in leukocytes. Biochim Biophys Acta 538:50–59.

Karnovsky MJ, Robinson JM, Briggs RT, Karnovsky ML (1981): Oxidative cytochemistry in phagocytosis: The interface between structure and function. Histochem J 13:1–22.

Karnovsky MJ, Kleinfeld AM, Hoover RL, Klausner RD (1982): The concept of lipid domains in membranes. J Cell Biol 94:1–6.

Karnovsky MJ, Kleinfeld AM, Hoover RL, Dawidowicz EA, McIntyre DE, Salzman EA, and Klausner RD (1983): Lipid domains in membranes. Ann NY Acad Sci 401:61–76.

Karnovsky ML, Wallach DFH (1961): The metabolic basis of phagocytosis. III. Incorporation of inorganic phosphate into various classes of phosphatides during phagocytosis. J Biol Chem 236:1895–1901.

Karnovsky ML (1983): Steps toward an understanding of Chronic Granulomatous Disease. N Engl J Med 308:274–275.

Kitagawa S, Takaku F, Sakamoto S (1980): Evidence that proteases are involved in superoxide production by human polymorphonuclear leukocytes and monocytes. J Clin Invest 65:74–81.

Klebanoff SJ, Clark RA (1978): The Neutrophil: Function and Clinical Disorders. Amsterdam: North Holland Publishing Co.

Nathan CF, Root RK (1977): Hydrogen peroxide release from mouse peritoneal macrophages. Dependence on sequential activation and triggering. J Exp Med 146:1648–1662.

Rossi F, Patriarca P, Romeo D (1980): Metabolic changes accompanying phagocytosis. In Sbarra AJ, Strauss RR (eds): "The Reticulo-Endothelial System: A Comprehensive Treatise. 2. Biochemistry and Metabolism." New York: Plenum Press, pp 153–158.

Segal AW, Cross AR, Garcia RC, Borregard N, Valerius NH, Soothill JF, Jones OTG (1983): Absence of cytochrome b_{-245} in chronic granulomatous disease: A multicenter European evaluation of its incidence and relevance. N Engl J Med 308:245–251.

Sklar LA, Miljanich GP, Dratz EA (1979): Phospholipid lateral phase separation and the partition of cis parinaric acid and trans parinaric acid among aqueous, solid lipid and fluid lipid phases. Biochemistry 18:1707–1716.

Yahara I, Harada F, Sekita S, Yoshihira K, Natori S (1982): Correlation between effects of 24 different cytochalasins on cellular structures and cellular events and those on actin in vitro. J Cell Biol 92:69–78.

III. WHITE CELL FLOW BEHAVIOR

White Cell Mechanics: Basic Science and
Clinical Aspects, pages 131–146

Radial Distribution of White Cells in Tube Flow

Harry L. Goldsmith and Samira Spain

McGill University Medical Clinic, Montreal General Hospital, Montreal,
Quebec, Canada H3G 1A4

INTRODUCTION

The appearance of white cells (WBC), notably granulocytes, at the periphery of vessels in the microcirculation is a well-documented phenomenon associated with low blood-flow states such as occur during inflammation [Grant, 1973; Zweifach, 1973]. The outward displacement of the white cells from the core of the bloodstream results in their creeping along the vessel wall, and is referred to as margination. This may be followed by cell-wall adhesion and subsequent emigration into the extravascular space, as first described by Dutrochet [1824]:

What we have just seen concerning the similarity of the organic composition of solids and fluids in the living body would indicate that the vesicular globules contained in the blood are added to the tissues of the organs and become fixed there to augment and repair them so that nutrition consists of a veritable intercalation of fully formed and extremely tiny cells. This opinion, though it may seem strange, is however well founded, since observation favors this view. Many times I have seen blood cells leaving the blood stream, being arrested and becoming fixed to the organic tissue. I have seen this phenomenon, which I was far from suspecting, when I observed the movement of the blood in the transparent tail of young tadpoles under the microscope. . . . Observing the movement of blood, I have seen many times a single cell escape laterally from the blood vessel and move in the transparent tissue ... with a slowness which contrasted strongly with the rapidity of the circulation from which the cell had escaped. Soon afterwards, the cell stopped moving and remained fixed in the transparent tissue. A comparison with the granulations which this tissue contained showed that they were in no way different. There is no doubt that these semitransparent granulations were also blood cells which had previously become fixed.

A number of investigators have since shown that white cell margination in whole blood is linked to the aggregation of red cells, which occurs at low flow rates. It is this aspect of the phenomenon which we have attempted to quantitate by measuring the distribution of white cells in straight uniform tubes of 100–180-μm diameter. Before describing the work, a brief discussion of the above-mentioned studies is given.

PREVIOUS STUDIES
Fahraeus Effect

Fahraeus [1929] and Vejlens [1938] were the first to undertake a quantitative study of white cell distribution in vitro and in vivo. According to these authors, white cells are carried mainly near the center of small vessels under normal, rapid conditions of flow. They ascribed the axial flow of WBC to their being larger than the red cells. To verify this hypothesis they measured the cell number concentration, n_T, in blood flowing through a 100-μm diameter tube and compared it with that in the feed reservoir, n_F, or in the discharge, n_D. Fahraeus [1929] had previously shown that, in the case of red cells, the ratio of tube to discharge hematocrit, $H_T/H_D < 1$ (the so-called Fahraeus effect). Provided that there is no screening effect of the cells entering the tube from the reservoir (Cokelet, 1976; Gaehtgens, 1980], at the steady state, mass balance demands that reservoir and discharge cellular concentrations be equal and that the ratio of tube to feed (or discharge) concentration be equal to the ratio of mean blood velocity $\overline{U}$, to mean cell velocity, $\overline{u}_{RBC}$ or $\overline{u}_{WBC}$:

$$\frac{n_T}{n_F} = \frac{n_T}{n_D} = \frac{\overline{U}}{\overline{u}_{RBC, WBC}}.$$

In the case of the red cells, $H_T/H_D = n_T/n_D < 1$ implies that $\overline{u}_{RBC} > \overline{U}$, and that there must therefore be a greater axial concentration of cells in the tube. Others have since shown that, provided the tube diameter < 30 μm, $H_F = H_D$ [Gaehtgens, 1980], $H_T/H_D < 1$, and the ratio decreases with decreasing tube diameter, down from 150 μm where the effect is first noticeable [Hochmuth and Davis, 1969; Barbee and Cokelet, 1971; Gaehtgens et al, 1978].

In the case of white cells, Vejlens reported an experiment in a 100-μm tube at a high flow rate which resulted in a mean $n_T/n_F = 0.75 \pm 0.07$ (SD), compared to a mean value of 0.82, previously obtained by Fahraeus for red cells under similar flow conditions. Thus, it was concluded that the larger WBC were more axially transported than the red blood cells (RBC). In fact, Vejlens was able to show that n_T/n_F for the lymphocytes (0.85 $\pm$ 0.13) was larger than that for the polymorphonuclear cells (0.71 $\pm$ 0.07), the mean diameters of these corpuscles being 6.20 and 7.25 μm, respectively [Schmid-Schönbein et al, 1980a].

Flow Rate, RBC Aggregation, and Margination of WBC

In vivo, Vejlens [1938] demonstrated that white cell margination was associated with red cell aggregation, which he induced by injecting gelatin or fibrinogen. The effect has since been studied in vitro by Palmer [1967], who measured the white cell concentration in samples collected at various transverse positions from blood flowing through a 30-μm wide rectangular slit. More recently, Nobis et al [1982] directly measured the distribution of fluorescent-labeled white cells in blood flowing through capillary tubes of 34–69-μm diameter. White cells migrated to the periphery of the tubes when red cell aggregation was induced either by reducing the flow rate or by adding high molecular weight dextrans.

Observations in the microcirculation by Palmer [1959] and Bagge [1975] indicate that, in periods of fast flow, fewer white cells pass through small capillaries than in periods of sluggish flow, suggesting that only at low flow rates are the white cells found at the periphery of precapillary vessels. An increase in the drift of white cells to the wall of vessels with decreasing flow rate has also been documented in 1-mm rabbit femoral arteries by rapid freezing of and sectioning the vessels [Phibbs, 1966]. In these arteries, even under normal physiological flow conditions, the concentration of white cells was found to increase toward the periphery.

Margination in Postcapillary Venules

A different mechanism, applicable to the margination of white cells principally observed in postcapillary venules has been proposed by Schmid-Schönbein et al [1980b]. Here, it was shown that, because of the interactions with the smaller, discoidal, and more deformable red cells, white cells in divergent flow emerging from a capillary ahead of a train of red cells are displaced toward the wall of the venule. The phenomenon was observed in the microcirculatory bed of the rabbit ear and subsequently modeled in vitro on a large scale using suspensions of macroscopic rigid spheres and flexible discs in glycerol flowing through divergent channels and sudden expansions of vessel lumen. Displacements of the spheres toward the wall by the discs, similar to those of the white cells by the red cells in post capillary venules, were seen.

WBC DISTRIBUTION IN GHOST CELL SUSPENSIONS

Initially, the distribution of tracer human white cells in thin-walled precision-bore glass tubes was obtained directly by observing them in transparent suspensions of reconstituted biconcave ghost cells [Goldsmith and Karino, 1977; Goldsmith and Marlow, 1979]. Using a hydraulically operated traveling microtube apparatus previously described [Goldsmith and Marlow, 1979; Takamura et al, 1979] the cells were photographed with a 16-mm cine camera through a microscope in the median plane of a vertically mounted tube as they flowed downward from a 150-μl infusion syringe through the glass tube, at rest. The cells were counted in five strips of equal width on either side of the tube axis, and the number concentration in each strip determined, taking into account the measured velocity distribution [Goldsmith and Spain, 1984].

The results showed that human white cells in native plasma migrated inward from the wall in flow down the tubes at mean linear flow rates $\overline{U} >$ 0.5 mm per second (s^{-1}), the rate of migration increasing with increasing flow rate. This is shown in the histogram of Figure 1a in which the normalized number concentration of cells is plotted against relative tube radius (radial distance from tube axis, R/tube radius, R_0).

By contrast, the results obtained in 45% ghost cell suspensions shown in Figure 1b at mean flow rates > 0.5 mm s^{-1} showed a drift of white cells toward the tube wall, the rate of outward migration increasing with increasing $\overline{U}$. This result was reminiscent of data previously obtained with red cells in ghost cell suspensions [Goldsmith and Marlow, 1979], as shown in Figure 2. There, the effect was attributed to the more rapid inward migration, due to inertia of the fluid, of ghost cells than red cells. The difference in migration rates is associated with a difference in the sign and magnitude of the sedimentation velocities [Brenner, 1966]. In plasma, based on the known density, size, and the Stokes equation, the sedimentation velocities are estimated to be ~ -0.1 μm s^{-1} for the ghost cells and $\sim +1$ μm s^{-1} for the red cells. A similar difference exists between the ghost cells and the white cells (from $\sim +0.7$ μm s^{-1} for lymphocytes to $\sim +2$ μm s^{-1} for granulocytes). The observed outward migration may therefore be an artefact of the ghost cell flow visualization technique and

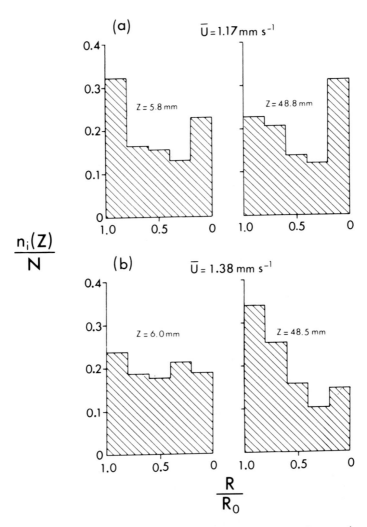

Fig. 1.   White cell distribution (a) in plasma, and (b) in a 45% ghost cell suspension within 110- and 180- μm diameter tubes, respectively. Histograms of the normalized cell number concentration across the median plane of the tube as a function of distance Z from the tube entry. Here, n_i (Z) is the number concentration in the ith strip of the median plane i = 1, 2 ... 10, and N is the number concentration in the whole tube. Significant changes in distribution with flow over distances Z > 40 mm were found at mean velocities $\overline{U}$ > 1 mm s^{-1} in plasma (inward migration of cells) and in ghost cells (outward migration). The data from the two sides of the median plane (i = 1 to 5, and 6 to 10) have been combined. Uniform cell distribution across the tube corresponds to $n_i(Z)/N$ = 0.2.

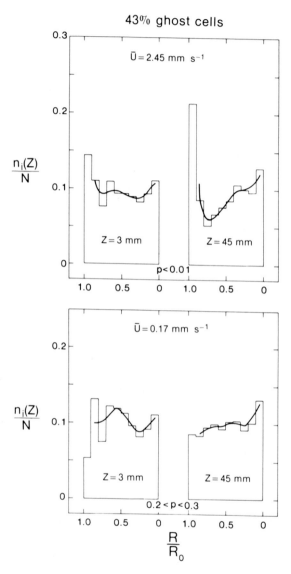

Fig. 2. Histograms as in Figure 1 of the normalized red cell distribution in a 43% ghost cell suspension within an 85-μm diameter tube. Significant outward migration of the red cells, indicated by the p values, occurred at high mean velocities $\bar{U} > 1$ mm s^{-1}. The solid lines were calculated by averaging between adjacent strips n_i $(Z) = 1/4$ $[2n_i$ $(Z) + n_{i-1}$ $(Z) + n_{i+1}$ $(Z)]$, since the radial position of cell centers could not be determined to better than 1 μm, the width of each strip being ~4 μm (from Goldsmith and Marlow [1979]).

would not be expected to occur in whole blood where there is no difference in sign in the sedimentation velocity.

WBC DISTRIBUTIONS IN WHOLE BLOOD: FAHRAEUS EFFECT

In whole blood flowing through tubes having diameters > 50 μm, the reflection and refraction of the transmitted light make it impossible to observe the motions of white cells in the core and to assess their number concentration. When the white cells in whole blood were stained with the fluorochrome acridine orange, and observed under UV illumination, it was visually evident that, at hematocrits of 20–30% and mean flow rates < 0.5 mm s^{-1} there was a higher concentration of WBC at the tube periphery. Unfortunately, the low intensity of the fluorescence made it impossible to record the cell motions on cine film and to measure their distribution within the tubes.

Instead, using the Fahraeus effect, the number concentration of white cells and red cells in the feed reservoir, n_F, and in 100- and 150-μm diameter tubes, n_T, were measured at reservoir hematocrits of 20%, 40%, and 60% [Goldsmith and Spain, 1984]. Blood flowed from a feed reservoir through vertically mounted sets of the 150- or 100-μm tubes into a mechanically operated withdrawal syringe at mean velocities $\overline{U}/2R_0$ from 9 to 140 tube diameters s^{-1}. This compares with a range of 0.3 to 18.9 diameters s^{-1} in a 99-μm tube, and a value of 5.1 diameters s^{-1} in a 154 μm tube, in the experiments of Barbee and Cokelet [1971]. After suddenly arresting the flow, the red cells and white cells in aliquots of the reservoir and in the blood eluted from the tubes were counted using a Coulter instrument.

Effect of flow rate. There was a marked increase in the white cell number concentration in the tube as the flow rate decreased to values of $\overline{U}/2R_0 < 20$ s^{-1}. The effect was more pronounced in the 100- than in the 150-μm tube, as illustrated in Figures 3 and 4 giving bar graphs of n_T/n_F at the highest and lowest $\overline{U}$. Thus, in the 100 μm tube, $n_T/n_F > 1.1$ at the lowest flow rate and at all hematocrits implying that $\overline{u}_{WBC} < \overline{U}$. At the highest flow rate, $n_T/n_F < 1$, and hence $\overline{u}_{WBC} > \overline{U}$.

In the case of the red cells, there was no such effect, in agreement with the results of Barbee and Cokelet [1971] and other investigators cited above, n_T/n_F being appreciably < 1 at all flow rates. Instead there was a small, though statistically insignificant, decrease in n_T/n_F in going from the highest to the lowest flow rate.

Although the white cell concentration in the tube exceeded that in the feed reservoir only at the lowest $\overline{U}$, there was an enrichment of white cells relative to red cells in both 100- and 150 μm tubes at higher $\overline{U}$ and at

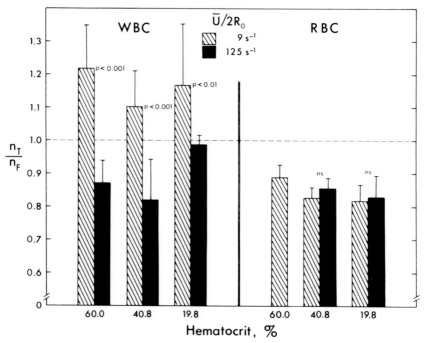

Fig. 3. Bar graphs of the number concentration of white and red cells in 100-μm tubes, n_T, relative to that in the feed reservoir, n_F, at high (hatched) and low (solid bars) mean velocities $\overline{U}$. At low $\overline{U}$, there was a significant increase in white cell tube concentration, which was not seen in the case of red cells.

nearly all hematocrits. This is shown in the upper part of Table I which gives values of the ratio of WBC:RBC concentration in the tube, $(n_{WBC}/n_{RBC})_T$, relative to that in the reservoir, $(n_{WBC}/n_{RBC})_F$. The enrichment of white cells in the tube, even at $\overline{U}/2R_0 > 25$ s^{-1} appears to be at variance with the results of Vejlens [1938], which showed that there was a depletion of white cells relative to red cells in the tube. In his experiment, blood at ~30% hematocrit (3 parts of venous blood to 1 part of citrate) was drawn through a 200-mm-long, 100-μm diameter tube at a negative pressure of 100 mm Hg. Assuming an apparent viscosity of 3 m Pascals (Pa s) at 23°C and applying the Poiseuille-Hagen equation yields $\overline{U}/2R_0 = 69$ s^{-1}. This would correspond to the upper half of the range in Table I where, at 40% hematocrit, the ratio $(n_{WBC}/n_{RBC})_T/(n_{WBC}/n_{RBC})_F$ decreased from 1.16 to 0.99 as $\overline{U}/2R_0$ increased from 30 to 125 s^{-1}. In the Vejlens experiment, the above ratio = 0.91 with $n_T/n_F = 0.75$ for the white cells.

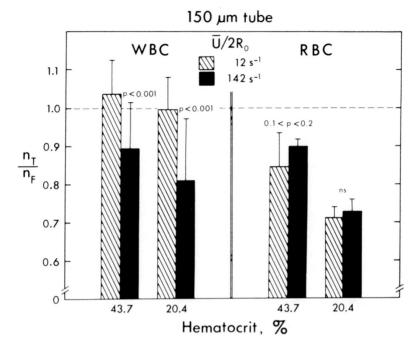

Fig. 4. Bar graphs, as in Figure 3 of the relative number concentration of white and red cells in 150-μm tubes. There is a less pronounced increase in white cell tube concentration at low $\overline{U}$.

Red Cell Aggregation Determines WBC Margination

The above results for the white cells imply that there was a decrease in the mean cell velocity with decreasing flow rate, and a redistribution of the cells toward the tube periphery. It was strongly suspected that the effect was directly related to red cell aggregation at low shear rate, since such aggregation when observed in tubes of ~ 100-μm diameter, is accompanied by inward migration of erythrocyte rouleaux [Meiselman, 1965; Merrill et al, 1965; Goldsmith, 1967]. The formation of a core of rouleaux could then displace white cells to the periphery thereby increasing n_T/n_F, as previously postulated by Nobis et al [1982] to explain their results in smaller glass tubes:

> . . . the radial frequency distribution of WBC in the flowing blood is not so much a consequence of their own flow behaviour or flow properties, but much more dependent on the interactions with the red blood cells surrounding them.

**TABLE I. White Cell: Red Cell Concentrations in
Tube and Reservoir**

$\overline{U}/2R_0$ (s^{-1})	Hematocrit (%)	$\dfrac{(n_{WBC}/n_{RBC})_T}{(n_{WBC}/n_{RBC})_F}$
150-μm tube, blood		
142	44.1	1.00 ± 0.17 (SD)
	19.0	1.10 ± 0.21 (SD)
12	43.1	1.25 ± 0.13 (SD)
	21.1	1.41 ± 0.11 (SD)
100-μm tube, blood		
125	41.3	0.99 ± 0.17 (SD)
	19.8	1.32 ± 0.19 (SD)
30	40.9	1.16 ± 0.16 (SD)
	20.3	1.29 ± 0.11 (SD)
9	60.4	1.33 ± 0.10 (SD)
	39.8	1.36 ± 0.13 (SD)
	19.7	1.50 ± 0.24 (SD)
150-μm tube, cells in phosphate buffer		
120	38.7	0.80 ± 0.07 (SD)
10	38.5	0.82 ± 0.07 (SD)
100-μm tube, cells in phosphate buffer		
116	39.6	0.82 ± 0.05 (SD)
9	40.1	0.87 ± 0.09 (SD)

To test the existence of such an effect at low $\overline{U}/2R_0$, cine films were taken in 100-μm tubes of blood at hematocrits of 20% and 39%. At mean flow rates < 10 tube diameters s^{-1} there was a two-phase flow of a well-defined core of red cell rouleaux with a few single cells, small rouleaux, white cells, and many platelets at the tube periphery. As the mean flow rate decreased further, the width of the core decreased and below 5 tube diameters s^{-1}, the core moved as a plug, surrounded by an almost red cell-free peripheral layer. As illustrated in Figure 5 at 20% hematocrit, the core width fluctuated with axial position in the tube, especially at low $\overline{U}$. The rapid decrease of the mean measure core width with decreasing flow rate, and at a given flow rate, with hematocrit is shown in Table II. At the lowest flow rate in the measurements of cell number concentrations, $\overline{U}/2R_0 = 9$ s^{-1}, the ratios of mean core width to tube diameter were 0.76 and 0.96 at 20% and 39% hematocrit, respectively, corresponding to a peripheral cell-depleted layer of ~12 and 2.5 μm, respectively.

WBC DISTRIBUTION IN WASHED BLOOD

Confirmation of the effect of red cell aggregation on the distribution of white cells in the tubes was sought in measurements of the number

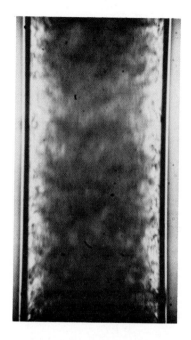 

$$\overline{U} = 1.54 \text{ mm s}^{-1} \qquad \overline{U} = 0.11 \text{ mm s}^{-1}$$

Fig. 5. Photomicrographs of the flow of blood at 20% hematocrit at high (left) and low (right) flow rates in a 100-μm tube. A core of rouleaux of red cells of nonuniform width develops at low $\overline{U}$. At the tube periphery, there are single rouleaux, platelets, and white cells.

concentration in washed cells suspended in isotonic phosphate buffer-albumin solution. In these suspensions, rouleau formation was totally suppressed. The results of experiments at 40% hematocrit are shown in Figure 6 and Table I (lower part).

The contrast between washed and whole blood is most striking. In washed blood the values of $n_T/n_F < 0.8$ and *lower* than those for the red cells at all flow rates. There was only a small and statistically insignificant increase in n_T/n_F for the white cells in going from high to low flow rates. Thus, as shown in Table I, there was a depletion of white cells relative to the red cells [$(n_{WBC}/n_{RBC})_T/(n_{WBC}/n_{RBC})_F < 1$], implying that $\overline{U}_{WBC} > \overline{u}_{RBC}$ and that therefore the white cells were, on average, being more axially transported than the red cells. The values of n_T/n_F for red cells in washed blood were only a little larger than those in whole blood.

TABLE II. Mean Relative Core Width of Red Cell Aggregates at Low Flow Rates (100 μm Tube)

Hematocrit (%)	$\overline{U}/2R_0$ (s^{-1})	Mean $\dfrac{\text{Core width}}{\text{Tube diameter}}$
Blood		
39.2	2.5	0.82 ± 0.09 (SD)[a]
	5.0	0.94 ± 0.03
	10.1	0.96 ± 0.03
20.4	3.4	0.61 ± 0.15
	4.8	0.68 ± 0.16
	9.7	0.79 ± 0.12
	25.6	0.93 ± 0.03
Cell in phosphate buffer		
19.8	10	0.99[b]
	81	0.99

[a]Distance between peripheries of outermost red cells of the aggregates constituting the core.

[b]No core of aggregated red cells; value represents distance between peripheries of the outermost red cells in the suspension.

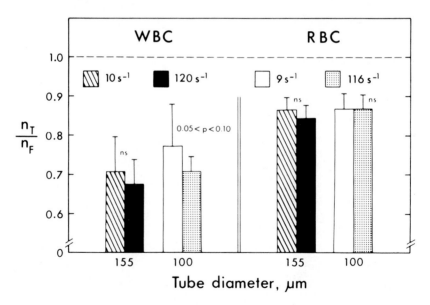

Fig. 6. Bar graphs, as in Figures 3 and 4, of the number concentration of cells in the tubes at 40% hematocrit in washed blood. By contrast to the earlier figures, the white cell relative number concentrations are ≤ 0.8, both at high and low $\overline{U}$, and are lower than the values for the red cells. The values of $\overline{U}/2R_0$ corresponding to the various bars are given in the upper part of the diagram.

Such a large reduction in white-cell concentration in the tube is difficult to explain on the basis of the known mechanisms for cell migration from the tube wall. Thus one can simply assume that the reduced concentration is due to exclusion of cell centers from a layer, one cell radius, b, in thickness at the tube wall, surrounding a core of increased but uniform concentration. Thomas [1962] has calculated values of n_T/n_F at a given wall shear stress, taking into account the non-Newtonian flow properties of the core of red cells:

$$\frac{n_T}{n_F} = \gamma^2 + \frac{(1 - \gamma^2)^2}{2(1 - \gamma^2) + \gamma^2 \phi_r},$$

where $\gamma = (R_0 - b)/R_0$ and ϕ_r is the relative fluidity in the core, a function of γ, n_T and the wall shear stress. If one now assumes that white cells are uniformly distributed with the red cells in the core, and that b = 3.5 μm ($\sim$ mean white cell radius), one can use known values of ϕ_r (eg, data of Barbee [1970], R_0 = 49.5 μm), and at 40% hematocrit calculate from the above equation that n_T/n_F = 0.95. The model therefore predicts a reduction in white cell concentration much smaller than that found.

This leaves the two known hydrodynamic mechanisms for radial migration of particles in the tube flow of dilute suspensions, both of which are wall effects. The first depends on particle deformation as modeled by the inward migration of liquid droplets [Karnis and Mason, 1967], and found to apply to red cells in viscous media [Goldsmith, 1971; Goldsmith and Mason, 1971]. Clearly, this mechanism would not be expected to apply to white cells in plasma or buffer solution, since deformation of a cell whose internal viscosity has recently been estimated to be 13 Pa s [Schmid-Schönbein et al, 1981], ie, $\sim$2,000 times that of the red cell, could not occur at shear stresses $\leq$ 1.2 Pa prevailing in the above-described experiments.

The second mechanism depends on inertia of the suspending fluid [Cox and Brenner, 1968] and results in a two-way migration, inward from the wall, and outward from the axis toward an eccentric equilibrium position [Segré and Silberberg, 1962]. Both rigid and deformable particles have been observed to undergo this two-way migration, provided the particle Reynolds number $Re_p = (8/3)b(b/R_0)^2 \overline{U}\rho/\eta > 10^{-4}$, b being the particle radius, ρ and η the fluid density and viscosity respectively. The tubular pinch effect has been demonstrated with normal and aldehyde-fixed red cells [Goldsmith, 1971; Goldsmith and Mason, 1971] and inward migration of these cells at high Re_p also observed in flow through a 25-μm-wide rectangular slit [Palmer and Betts, 1975]. The rate of migration in dilute suspensions is found to increase as $(b/R_0)^3$ [Segré and Silberberg, 1962; Karnis et al, 1966]. One would therefore expect inward migration of white cells (b = 3.5 μm) at the highest flow rates ($Re_p > 10^{-4}$) to occur at velocities about twice those of the red cells, assuming an equivalent sphere radius of ~2.7 μm for the latter. However, this ignores the important outward dispersive effect of cell interactions at normal hematocrits. More important, it is still necessary to account for the remarkably low value of n_T/n_F for white cells in washed blood at the lowest $\overline{U}/2R_0 = 9$ s^{-1} when $Re_p = 4 \times 10^{-5}$ and there should be little inward migration, even in dilute suspensions [Goldsmith and Spain, 1984].

CONCLUDING REMARKS

Outward displacement of white cells occurs at low flow rates when there is appreciable aggregation of red cells, accompanied by inward migration of red cell rouleaux, the latter being responsible for the outward movement of white cells. No such effect is observed either at high flow rates, when the fluid shear stresses are of sufficient magnitude to break up red cell aggregates, or in washed blood, when there is no red cell aggregation. That outward margination is not observed in the ghost cell system at low flow rates is presumably due to the fact that large aggregates do not form at low shear rates in this system.

The finding that there is a relative enrichment of white cells: red cells even at values of $\overline{U}/2R_0 > 25$ s^{-1}, corresponding to mean tube shear rates (based on Poiseuille flow) > 100 s^{-1}, at which red cell rouleaux are known to be broken up, would indicate some outward displacement of white cells, presumably from a central core of small radius where local shear rates are low enough to permit aggregates to form. In this regard, the finding of Phibbs [1966] in 1-mm diameter rabbit femoral arteries, that white cells are peripherally concentrated, may be significant.

ACKNOWLEDGMENTS

This work was supported by grant MT-1835 from the Medical Research Council of Canada and a grant from the Quebec Heart Foundation. H.L.G. is a Career Investigator of the Medical Research Council of Canada. We gratefully acknowledge the assistance of the above granting agencies, and wish to thank Diane Chajczyk and Claire Goldsmith for their technical assistance and Pamela Lilley for typing the manuscript.

REFERENCES

Bagge U (1975): White blood cell rheology. Experimental studies on the rheological properties of white blood cells in man and rabbit and in an *in vitro* micro-flow system. Dissertation, University of Göteborg.

Barbee JH (1970): The flow of human blood through capillary tubes with inside diameters between 8.7 and 221 microns. PhD thesis, California Institute of Technology.

Barbee JH, Cokelet GR (1971): The Fahraeus effect. Microvasc Res 3:6–16.

Brenner H (1966): Hydrodynamic resistance of particles at small Reynolds numbers. In Drew TB, Hoopes JW, Vermeulen T (eds): "Advances in Chemical Engineering," Vol 6. New York: Academic Press, pp 287–438.

Cokelet GR (1976): Macroscopic rheology and the tube flow of human blood. In Grayson J, Zingg W (eds): "Microcirculation," Vol 1. New York: Plenum Press, pp 9–31.

Cox RG, Brenner H (1968): The lateral migration of solid particles in Poiseuille flow. I. Theory. Chem Eng Sci 23:147–173.

Dutrochet MH (1824): Recherches anatomiques et physiologiques sur la structure intime des animaux et des végétaux, et sur leur motilité. Paris: Baillière et Fils.

Fahraeus R (1929): The suspension stability of blood. Physiol Rev 9:241–274.

Gaehtgens P (1980): Flow of blood through narrow capillaries: Rheological mechanisms determining capillary hematocrit and apparent viscosity. Biorheology 17:183–189.

Gaehtgens P, Albrecht KH, Kreutz F (1978): Fahraeus effect and cell screening during tube flow of blood. I. Effect of variation of flow rate. Biorheology 15:147–154.

Goldsmith HL (1967): Microscopic flow properties of red cells. Fed Proc 26:1813–1820.

Goldsmith HL (1971): Red cell motions and wall interactions in tube flow. Fed Proc 30:1578–1588.

Goldsmith HL, Karino T (1977): Microscopic considerations: The motions of individual particles. Ann NY Acad Sci 283:241–255.

Goldsmith HL, Marlow JC (1979): Flow behavior of erythrocytes. II. Particle motions in concentrated suspensions of ghost cells. J Colloid Interface Sci 71:383–407.

Goldsmith HL, Mason SG (1971): Some model experiments in hemodynamics. IV. In Hartert HH, Copley AL (eds): "Theoretical and Clinical Hemorheology." New York: Springer, pp 47–59.

Goldsmith HL, Spain S (1984): Margination of leukocytes in blood flow through small tubes. Microvasc Res (in press).

Grant L (1973): The sticking and emigration of white blood cells in inflammation. In Zweifach BW, Grant L, McCluskey RT (eds): "The Inflammatory Process," Vol II. New York: Academic Press, pp 205–249.

Hochmuth RM, Davis DO (1969): Changes in hematocrit for blood flow in narrow tubes. Bibl Anat 10:59–65.

Karnis A, Goldsmith HL, Mason SG (1966): The flow of suspensions through tubes. V. Inertial effects. Can J Chem Eng 44:181–193.

Karnis A, Mason SG (1967): Particle motions in sheared suspensions. XXIII. Wall migration of fluid drops. J Colloid Interface Sci 24:164–169.

Meiselman HJ (1965): Some physical and rheological properties of human blood. ScD thesis, Massachusetts Institute of Technology.

Merrill EW, Benis AM, Gilliland ER, Sherwood TK, Salzman EW (1965): Pressure-flow relations in human blood in hollow fibers at low flow rates. J Appl Physiol 20:954–967.

Nobis U, Fries AR, Gaehtgens P (1982): Rheological mechanisms contributing to WBC-margination. In Bagge U, Born GVR, Gaehtgens, P (eds): "White Blood Cells:

Morphology and Rheology as Related to Function." The Hague/Boston: Martinus Nijhoff, pp 57–65.

Palmer AA (1959): A study of blood flow in minute vessels of the pancreatic region of the rat with reference to intermittent corpuscular flow in individual capillaries. Quart J Exp Physiol 44:149–159.

Palmer AA (1967): Platelet and leukocyte skimming. Bibl Anat 9:300–303.

Palmer AA, Betts WH (1975): The axial drift of fresh and acetaldehyde-hardened erythrocytes in 25 μm capillary slits of various lengths. Biorheology 12:283–291.

Phibbs RH (1966): Distribution of leukocytes in blood flowing through arteries. Am J Physiol 210:919–925.

Schmid-Schönbein GW, Shih YY, Chien S (1980a): Morphometry of human leukocytes. Blood 56:866–875.

Schmid-Schönbein GW, Usami S, Skalak R, Chien S (1980b): The interaction of leukocytes and erythrocytes in capillary and post-capillary vessels. Microvasc Res 19:45–70.

Schmid-Schönbein GW, Sung K-L, Tozeren H, Skalak R, Chien S (1981): Passive mechanical properties of human leukocytes. Biophys J 36:243–256.

Segré G, Silberberg A (1962): Behaviour of macroscopic rigid spheres in Poiseuille flow. II. Experimental results and interpretation. J Fluid Mech 14:136–157.

Takamura K, Goldsmith HL, Mason SG (1979): The microrheology of colloidal dispersions. IX. Effects of simple and polyelectrolytes on rotation of doublets of spheres. J Colloid Interface Sci 72:385-400.

Thomas HW (1962): The wall effect in capillary instruments: An improved analysis suitable for application to blood and other particulate suspensions. Biorheology 1:45–56.

Vejlens G (1938): The distribution of leukocytes in the vascular system. Acta Pathol Microbiol Scand [Suppl] 33:11–239.

Zweifach BW (1973): Microvascular aspects of tissue injury. In Zweifach BW, Grant L, McCluskey RT (eds): "The Inflammatory Process," Vol II. New York: Academic Press, pp 3–46.

White Cell Mechanics: Basic Science and
Clinical Aspects, pages 147–157
© 1984 Alan R. Liss, Inc., 150 Fifth Avenue, New York, NY 10011

Flow Behaviour of White Cells in Capillaries

P. Gaehtgens, A. R. Pries, and U. Nobis

Institut für Physiologie der Freien Universität Berlin, 1000 Berlin 33 (P.G., A.R.P.) and Institut für Normale und Pathologische Physiologie der Universität, 5000 Köln-Lindenthal (U.N.), West Germany

INTRODUCTION

Capillary blood vessels differ greatly in luminal dimensions. Therefore, the behaviour of circulating white blood cells (WBCs) and their effect on flow of other blood constituents through individual capillaries will be variable, depending on the actual diameter of the vessel (D) and that of the white cell (d).

In theory, two situations can be distinguished:

1. A capillary vessel which is large enough to allow passage of a spherical white cell without necessitating cellular deformation (d/D < 1.0). In this case, the flow characteristics will mainly be affected by interactions between the WBC and other blood cells.

2. A capillary vessel which is sufficiently narrow to allow WBC passage only after cellular deformation from the spherical shape (d/D > 1.0). In this case, the micromechanical properties of the WBC itself may become more important than its interaction with other blood constituents.

In reality, both situations may occur in a single capillary: Capillary vessels may exhibit a local stenosis, often close to their entrance (eg, due to endothelial cell bulging which forces the WBC to transiently stop and gradually deform), but may for most of their length allow free passage of a WBC.

In this communication some characteristic effects caused by WBCs in the flow of blood through narrow vessels will be discussed. This analysis is based on experiments performed in vitro; such an approach allowed determination or manipulation of parameters which are more difficult (or impossible) to obtain or control in vivo because of the geometric complexity and the hemodynamic instability of the capillary circulation.

METHODS AND MATERIALS

The experiments reported here were performed with human venous blood, anticoagulated with edetic acid (EDTA) (2.5 mg/ml). If required, red cell-free suspensions of WBCs in plasma (containing also platelets) were obtained by allowing spontaneous sedimentation of the red cells (RBC) in a 37°C water bath.

Cell suspensions were perfused at constant but variable pressure through glass capillary tubes with various diameters. The flowing cells were observed through a microscope and their image recorded on videotape using a low-light-level TV-camera. WBCs were visualized by fluorescence microscopy after labelling with acridin orange (0.01 g/ml). The travelling capillary method [Gaehtgens et al, 1980] was employed to study WBC/RBC-interactions. Cell velocities were determined off-line using frame-by-frame-analysis. Quantitative densitometry was utilized according to the technique described by Pries et al [1983] to determine intracapillary hematocrit.

RESULTS AND DISCUSSION

Train Formation

Undeformed (spherical) WBCs differ both in volume and in deformability from the many times more numerous red cells. In the typical case (1), the linear velocity of the WBC (^{v}WBC) is usually lower than that of the red cells (^{v}RBC). As a consequence of this velocity difference, red cells preceding the WBC will move away from it (causing formation of a "plasma gap"), while those following the WBC will seize up, causing a typical flow pattern called "train-flow." Train-flow is consistently formed in vessels (or tubes) the diameter of which is less than approximately 15 μm. Train-flow is, however, also found in vessels with larger diameters downstream of a stenosis: Figure 1 shows the relative frequency of train-forming WBCs upstream (on the "arterial" side) and downstream (on the "venous" side) of a 12-μm stenosis in a glass tube with an inner diameter of 26 μm, perfused with human whole blood.

In a tube of given dimensions, the extent of train formation must depend on the velocity ratio between red and white cells and on the time required for cell transit through the capillary. If the capillary is sufficiently wide, single RBCs in the train overtake the WBC which is displaced to an off-center position [Schmid-Schönbein et al, 1980; Nobis

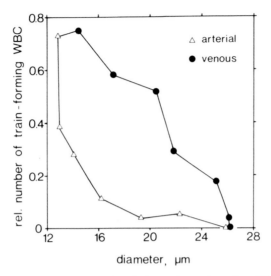

Fig. 1.   Number of train-forming WBCs relative to the total number of WBCs observed in a 26-μm glass capillary exhibiting a local stenosis of 12 μm. "Arterial" refers to the prestenotic section, "venous" to the poststenotic section of the capillary. The capillary was perfused with human whole blood, leukocytes were visualized by fluorescence labelling, and combined epi- and transillumination was employed.

and Gaehtgens, 1981]. In this case, the velocity of the single RBC after passage (ie, after axial reorientation) can be measured and compared to that of the WBC [Nobis and Gaehtgens, 1981]. Such data can also be obtained from densitometric recordings made on tubes with varying diameters. An example is shown in Figure 2 which demonstrates recordings made during perfusion of glass capillaries with human whole blood (hematocrit 0.45). An increased light transmission is recorded during passage of the plasma gap across the densitometric window, followed by a decrease in light transmission during passage of the high-hematocrit train of red cells. Calculation of the velocity ratio between red and white cells from the relative lengths of the plasma gap and the train yields values ranging between 1.0 and 1.3.

The length of the red cell train is obviously a function of the distance between tube orifice and the point of observation. Preliminary measurements have confirmed the expected linear increase of train length with the distance travelled by the train. Train length at any point of observation is, however, also a function of the individual velocity ratio ^{v}RBC/^{v}WBC. As shown in Figure 3, the train length increases with ^{v}RBC/^{v}WBC at low velocity ratios, but tends to decrease again at higher velocity ratios.

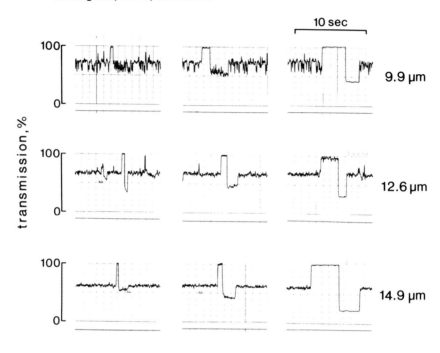

Fig. 2. Videophotometric recordings of light transmission through blood-perfused glass capillaries with the diameters indicated on the right. Typical patterns resulting from the effect of WBCs on red cell flow are seen: increased light transmission during passage of the plasma gap preceding white cell and decreased transmission during passage of red cell train.

Recordings such as those shown in Figure 2 can also be used to calculate the hematocrit in the train, if the velocity ratio $^vRBC/^vWBC$ and the "baseline hematocrit" are known. Results of such calculations are shown in Figure 4. In any capillary tube studied, the train hematocrit increases linearly with the velocity ratio $^vRBC/^vWBC$, and extremely high hematocrits may be reached, depending on the level of the "baseline hematocrit." Since the flow velocity imposed on the RBCs by the slow WBC is lower than their normal velocity, an increased "dynamic" RBC concentration behind the WBC would be expected. Although some "forward leak" of plasma past the WBC must be taking place, and even red cells may be "swept" past the WBC (in a forward direction) in the process (Fig. 5), the experimentally observed concentration of red cells behind the WBC is sufficiently explained by the local elevation of tube hematocrit resulting from a reduced Fahraeus effect of the red cells.

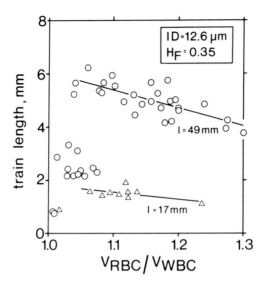

Fig. 3.   Train length as a function of velocity ratio between red and white blood cells in a 12.6-μm capillary perfused with whole blood at a feed hematocrit (H_F) of 0.35. Length of trains were determined at two points of observation, 49 and 17 mm downstream of the tube orifice.

Effect of WBC on Flow Resistance

The presence of WBCs in the flow of blood through capillaries may also influence the resistance to flow encountered. It is difficult to separate the effect on resistance due to the relatively rigid WBC per se from the possible additional effect caused by the red cell train following it. These two components can, however, be differentiated by comparing the flow behaviour of the white cell in the presence, with that in the absence, of red cells. If the velocity of a single WBC is measured at various positions along the length of a stenosed capillary, it may be expected that a reduction of velocity occurs when the cell arrives at a sufficiently narrow section of the capillary, where interaction with the wall occurs. Depending on the diameter of the stenosis, the velocity in the tapered section leading into the stenosis might in some way reflect the cell's ability to deform in response to the forces acting upon it. By comparing the actually observed WBC velocity with the velocity of an arbitrary fluid volume (which must accelerate in a predictable way as the tube cross-section decreases and vice versa), the effect on resistance caused by the WBC can be evaluated.

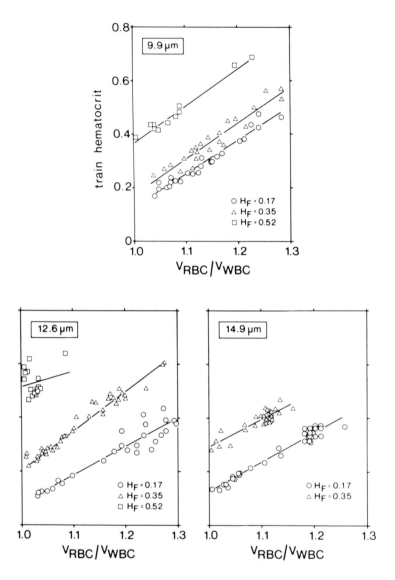

Fig. 4.   Train hematocrit as a function of velocity ratio between red and white blood cells in three capillary tubes. Diameters and feed hematocrits of the blood are indicated. Train hematocrit increases linearly with velocity ratio. Extrapolation of regression line to ʳRBC/ʳWBC = 1.0 gives the "baseline" tube hematocrit in the absence of WBCs.

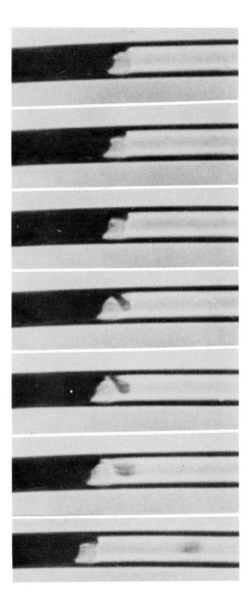

Fig. 5. Train flow in a 12-μm capillary tube. A single RBC is swept past the leading WBC in an off-center position and after passage is reoriented in the center-stream. Red cell train appears black due to observation at 420-nm wavelength.

Figure 6 shows the normalized fluid velocity in a capillary of 16-μm internal diameter (ID) with a stenosis of 4 μm, and, in addition, the velocity of several WBCs as a function of their instantaneous distance from the narrowest portion of the tube. In this experiment the stenosed capillary was perfused with a suspension of WBCs in their native plasma, obtained by spontaneous sedimentation of an anticoagulated (EDTA)

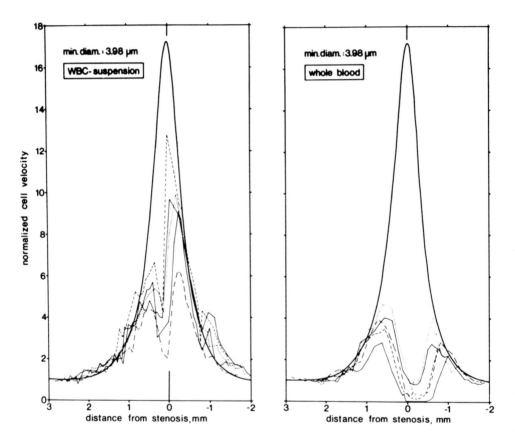

Fig. 6. Velocity of single WBCs as a function of instantaneous distance from the narrowest section of a stenosed capillary tube. Positive values of distance are prestenotic, negative values poststenotic. Minimal diameters of the tube was 3.98 μm, maximum diameters (before and after stenosis) was 16 μm. In addition, the calculated velocity of an arbitrary fluid element is given (solid line); all velocities are normalized with respect to the velocity in the straight section of the capillary tube. Left. Data obtained with WBCs suspended in their native plasma. Right. Data obtained during perfusion of whole blood.

venous blood sample. The cell velocity was determined from videorecordings of fluorescence-labelled white cells. As the cells approach the tapered part of the tube, their velocity increases as expected, and after reaching a diameter of approximately 6–7 μm, slows down more or less suddenly. The lowest velocity is seen in the narrowest section but a complete stop of motion is not observed. Downstream of the stenosis, the cells speed up again and finally return to their original velocity. Although differences in individual behaviour can be observed which probably reflect differences in single cell size and deformability, the general pattern is relatively uniform.

Figure 6 also shows data obtained during perfusion of the same stenosed capillary with human whole blood. Again, only the velocity of the WBCs was determined as a function of their instantaneous distance from the stenosis. While a qualitatively similar pattern of velocity is observed in this case, significant quantitative differences are also apparent: In all cases the maximum velocity is lower than that seen in the absence of RBC, and almost complete flow stop (even if only transient) is observed in the stenosis section of the capillary. Interestingly, the velocity of all WBCs studied was consistently *lower* downstream of the stenosis compared to that in an upstream section with the same diameter.

These data can be used to obtain some parameter equivalent to flow resistance by dividing the fluid velocity at any point along the length of the capillary by the corresponding WBC velocity. This "relative resistance" is plotted in Figure 7 for both sets of experiments. The comparison clearly shows that in the absence of RBCs the resistance increases up to maximally tenfold, when the white cell reaches the narrowest portion of the capillary, while it increases much more if red cells are also present in the perfusate. Furthermore, the resistance is lower downstream than upstream of the stenosis in the absence of red cells, while it is significantly higher in their presence.

These observations are interpreted as follows: The resistance to flow caused by a single white cell is noticeably increased when the capillary diameter decreases below approximately 6 μm. Lower resistances downstream of the stenosis compared to upstream indicate that the time required for restoration of the undeformed cell shape is very long compared to the transit time of the cell through the capillary. The major effect on flow resistance of a single WBC, however, results from its influence on red cell flow: The contribution of the RBC train to resistance may be significantly higher than that of the leukocyte alone; this is also the explanation of the low downstream velocity of the WBCs which is due to the presence of the train in the stenosis.

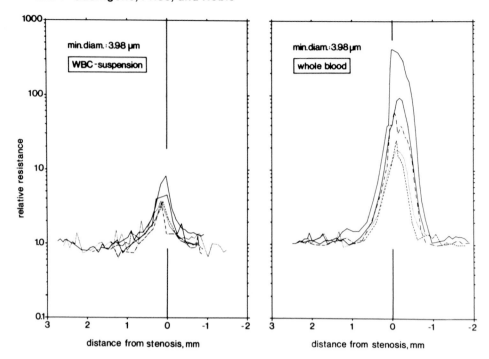

Fig. 7. "Relative resistance" calculated as the ratio of normalized fluid velocity and observed WBC velocity as a function of distance from the stenosis. Same experiment as in Figure 6. Note that prestenotic resistance is slightly higher than poststenotic resistance in the absence of RBC, while in their presence resistance is maximal only after the WBC has already passed the narrowest section of the capillary.

CONCLUSIONS

While these observations in vitro help to evaluate the effect of WBCs on blood flow in capillaries, they should not be applied without caution to the living capillary circulation. In general, the microvessels do not show the tapered geometry present in the in vitro model, and diameter changes tend to occur more abruptly at microvascular bifurcations. Furthermore, stenotic sections within single capillaries in vivo tend to be rather short, particularly if caused by endothelial cell bulging. In this case, the micromechanical properties of the WBCs certainly play a dominant role, even if red cells are also present in the capillary. This conclusion is supported by the observation of in vivo capillary plugging by WBCs, which in most cases occurs at distinct sites along the length of a single capillary. Such local stenotic sites are often shorter in length than the deformed white cell. Finally, the train phenomenon described here will probably be less pronounced in vivo under normal circumstances because vessel lengths are short relative to flow velocity and frequent branching of vessels will help to separate the train of red cells from the leading white cells. The phenomena described will, however, be relevant in longer capillary vessels and at reduced driving pressures which result in prolonged transit times.

ACKNOWLEDGMENTS

This study was supported by the Deutsche Forschungsgemeinschaft.

REFERENCES

Gaehtgens P, Dührssen C, Albrecht KH (1980): Motion, deformation, and interaction of blood cells and plasma during flow through narrow capillaries. Blood Cells 6:799–812.
Nobis U, Gaehtgens P (1981): Rheology of white blood cells during blood flow through narrow tubes. Bibl Anat 20:211–214.
Pries AR, Kanzow G, Gaehtgens P (1983): Microphotometric determination of hematocrit in small vessels. Am J Physiol 245:4167–4177.
Schmid-Schönbein GW, Usami S, Skalak R, Chien S (1980): The interaction of leukocytes and erythrocytes in capillary and post-capillary vessels. Microvasc Res 19:45–70.

White Cell Mechanics: Basic Science and
Clinical Aspects, pages 159–165

Deformation and Activation of Leukocytes—Two Contradictory Phenomena?

P. Gaehtgens

Institut für Physiologie der Freien Universität Berlin, 1000 Berlin 33, West Germany

It is well documented already in the older literature that active emigration of granulocytes from the bloodstream into the extravascular tissue occurs only in postcapillary venules, but not in true capillaries. This may seem surprising, since the geometrical conditions in narrow capillaries would appear to favour biochemical interaction between the membrane of a leukocyte and the endothelial surface. Nevertheless, granulocyte activation does not occur in capillaries, at least not in the sense of pseudopod formation eventually leading to active locomotion through the capillary wall.

The purpose of this contribution is to propose the hypothesis that pseudopod formation of granulocytes is inhibited as long as the cell is passively deformed. Due to geometrical constraint in narrow capillaries a leukocyte must change shape before passing the vessel with the bloodstream. Observations both in vivo and in vitro [Bagge et al, 1977a, b] show that this is a slow process resulting in a more or less cylindrical cell shape. Morphological studies [Schmid-Schönbein et al, 1980] have demonstrated that in the passive and nondeformed state of the cell (ie, in the spherical shape) the membrane is ruffled and exhibits folds (Fig. 1), indicating that the membrane material present is in excess of that needed to envelop the cell's volume: Published data suggest an excess membrane area between 90% [Schmid-Schönbein et al, 1980] and 40% [Bagge, 1976].

During deformation in capillaries, this "excess membrane," or at least some part of it, must inevitably be recruited and membrane folds must

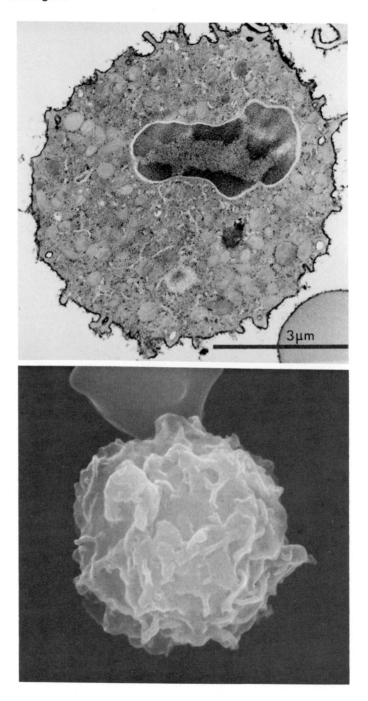

3μm

gradually disappear, depending on the extent of deformation. Electron microscopy of granulocytes trapped in narrow capillaries (Fig. 2) indeed give the impression of a rather smooth membrane [Bagge, 1976]. Although the extent of membrane recruitment during deformation of white cells in narrow capillaries has not been quantitatively studied, simple calculation assuming a spherical and a cylindrical cell shape at equal cell volume will yield the relationship shown in Figure 3: As the diameter of the capillary and thus of the deformed white cell decreases, the increased surface area relative to that of the sphere becomes more and more apparent. If an excess membrane of approximately 90% in the spherical state [Schmid-Schönbein et al, 1980] is assumed, this will be totally recruited in a capillary of approximately 2.7 μm diameter; if the excess membrane is 40% [Bagge, 1976], the minimum cylindrical diameter will be 3.7 μm.

On the other hand, membrane material must also be recruited during the process of pseudopod formation, since any irregular shape of the cell requires a large surface area: volume ratio compared to the sphere. Again, no quantitative data seem to be available on membrane "unruffling" during pseudopod formation, but the extent of membrane recruitment will clearly be a function of the number and dimensions of the pseudopods. Figure 4 shows a granulocyte during active emigration through the wall of a bone marrow sinus. The cell shape is rather irregular, but membrane ruffling seems to be less pronounced than in the resting cell.

It does not appear unreasonable to conclude from these considerations that the phenomena of deformation and pseudopod formation may not occur independently. With some degree of overstatement, the hypothesis presented here would predict that a sufficiently deformed granulocyte cannot be activated to form pseudopods and will therefore not be capable of active locomotion.

Specific experiments to verify or discredit this hypothesis seem not to have been performed so far. However, the above-mentioned in vivo observations would be consistent with this hypothesis, and the nonemigration of granulocytes from capillary vessels even after longer periods of "trapping" would be rather simply explained. An inhibition of pseudopod formation due to membrane "unruffling" in the deformed cell could be counteracted if membrane material could in sufficient amounts be recruited from intracellular sources. Studies in amoebae indicate that the turnover of

---

Fig. 1. Transmission (top) and scanning (bottom) electron micrograph of human neutrophils in the resting state, showing pronounced ruffles and folds of the membrane while the cell is in an approximately spherical shape (courtesy of Dr. GW Schmid-Schönbein, La Jolla, CA).

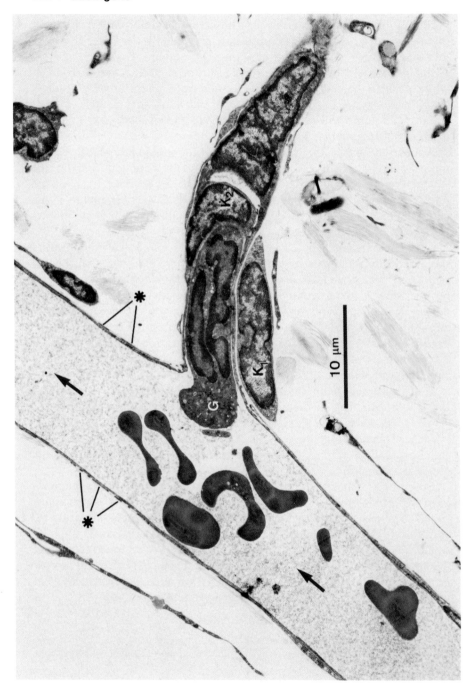

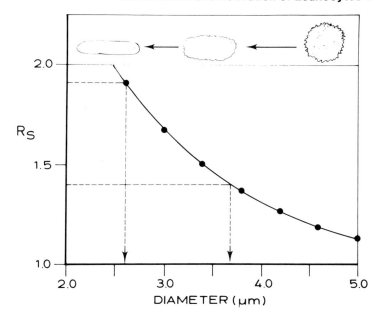

Fig. 3. Ratio (R_s) of the surface area of the deformed cell (in the shape of a cylinder with two hemispherical caps at the end) to that of a sphere of equal volume (200 μm^3) as a function of the diameter of the deformed cell. The arrows indicate the deformed cell diameter at which the excess membrane (90% according to Schmid-Schönbein et al [1980]; 40% according to Bagge [1976]) is fully recruited.

membrane material associated with endocytosis is approximately 0.2% of total membrane surface per minute; this may increase up to approximately 2% per minute by induced endocytosis for time periods not exceeding 15–30 minutes [Stockem, 1973]. Furthermore, no correlation was found between the rate of locomotion and endocytosis [Komnick et al, 1972]. Information on the possible recruitment of membrane material in granulocytes upon pseudopod formation (eg, by degranulation) does not appear to be available. Experimental studies attempting to induce pseudopod formation in the deformed granulocyte would serve to prove or disprove the validity of the hypothesis presented here.

Fig. 2. Electron micrograph showing granulocyte (G) trapped in the entrance section of a mesenteric capillary. Arrows indicate flow direction in the feeding arteriole, ascertained by intravital microscopy. K_1 and K_2 represent nuclei of a pericyte and a capillary endothelial cell, respectively. Note the thin but uninterrupted gap between granulocyte membrane and endothelium. Membrane "ruffles" comparable to those seen in Figure 1 are not observed. (Courtesy of Prof. Dr. F. Hammersen, München, Germany.)

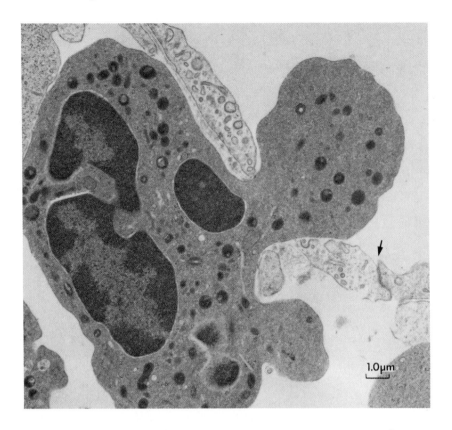

Fig. 4. Granulocyte in the process of active penetration through the wall of a marrow sinus. The arrow is in the sinus lumen and points to endothelial cell junction. Granulocyte shape is irregular and membrane ruffling is absent (from Lichtman [1982], by permission of the author).

It may also be worthwhile to point out that the process of deformation of a white cell, but even more the restoration of its spherical resting shape *after* deformation is a relatively slow process, certainly in comparison to the transit time of a cell through a single capillary. According to our own in vitro experience and to data in the literature [Bagge et al, 1977b] shape restoration may take tens of seconds or even minutes, depending on the length of time during which the cell was kept in the deformed state. If these data are applied to the living microcirculation, the leukocytes emerging from a narrow capillary would not be able to return to their original spherical shape before being carried by the flow all the way through the entire system of small venules (and possibly even further).

Therefore, pseudopod formation and consequently emigration would be hampered even in postcapillary venules, unless the cells are "caught" by an adhesive mechanism on the endothelial surface. Only hereby would sufficient time be granted to return to a geometrical configuration in which activation could ensue.

No systematic study of the shape recovery of granulocytes after capillary transit in vivo seems available from the literature. Such a study may, however, be warranted in order to elucidate the interaction of the various mechanical and biochemical phenomena which play a role in leukocyte behaviour in the microcirculation.

REFERENCES

Bagge U (1976): Granulocyte rheology. Blood Cells 2:481–490.

Bagge U, Johansson BR, Olofsson J (1977a): Deformation of white blood cells in capillaries. Adv Microcirc 7:18–28.

Bagge U, Skalak R, Attefors R (1977b): Granulocyte rheology. Experimental studies in an in vitro micro-flow system. Adv Microcirc 7:29–48.

Komnick H, Stockem W, Wohlfarth-Bottermann KE (1972): Ursachen, Begleitphänomene und Steuerung zellulärer Bewegungserscheinungen. Fortschr Zool 21:1–74.

Lichtman MA (1982): The anatomic features of leukocyte egress into the narrow sinus. In Bagge U, Born GVR, Gaehtgens P (eds): "White Blood Cells, Morphology and Rheology as Related to Function." The Hague, Boston, London: M.Nijhoff Publ, pp 69–77.

Schmid-Schönbein GW, Shih YY, Chien S (1980): Morphometry of human leukocytes. Blood 56:866–875.

Stockem W (1973): Pinozytose und Bewegung von Amoeben. IV. Quantitative Untersuchungen zur permanenten und induzierten Endocytose von Amoeba proteus. Z Zellforsch 136:433–446.

IV. WHITE CELL LOCOMOTION AND INTERACTION WITH ENDOTHELIUM

White Cell Mechanics: Basic Science and
Clinical Aspects, pages 169–193
© 1984 Alan R. Liss, Inc., 150 Fifth Avenue, New York, NY 10011

Chemotaxis, Chemokinesis, and Movement

Anthony T. W. Cheung and Michael E. Miller

Department of Pediatrics, School of Medicine, and California Primate Research Center, University of California at Davis, Davis, California 95616

The phenomenon of human polymorphonuclear leukocyte (PMN) movement and the mechanics of this movement have fascinated scientists for decades [Metchnikoff, 1893]. The ability of PMNs to orient and move rapidly enables them to arrive at the site of injury or infection as a "first line of defense" against microbial invaders. Upon entering the inflammatory arena, normal PMNs are then able to ingest (phagocytize) and kill many types of microorganisms. Among the cellular activities typical of an inflammatory response, movement is considered one of the most crucial functional characteristics of PMNs. In recent years, the recognition of intrinsic disorders of PMN movement has led to renewed interest in this field [Clark, 1978; Miller, 1975; Snyderman and Pike, 1977]. In order to fully understand the various perturbations of human PMN movement, it will be necessary to characterize the normal functional characteristics of PMN movement.

A number of cellular events has been identified to associate with PMN motility. In the presence of a chemoattractant gradient, PMNs show increased adherence and aggregation [Mease et al, 1980; Miller and Stiehm, 1981]. Movement of surface receptors to a localized area of the cell surface (capping) has also been shown among the cellular activities [Kimura et al, 1981]. Along with other cellular events, our interest has focused upon chemotaxis, chemokinesis, deformability and their relationship to PMN movement.

For many years, the term "Chemotaxis" was applied to the general phenomenon of a PMN moving toward a chemical gradient. In recent terminology, chemotaxis refers to the direction or orientation of locomo-

TABLE I. Glossary

Chemoattractant	A chemical or physiologically active substance which can trigger a chemotactic response in PMNs.
Chemokinesis	The enhanced motile activity of PMNs in terms of cellular deformation and translocation.
Chemotaxis	The orientation of PMNs to an established chemoattractant gradient.
Gradient slide	A Plexiglas chamber-slide capable of setting up and maintaining a reliable chemoattractant gradient (see Appendix for details).
Orientation	Directional alignment of PMNs during a chemotactic response.
Oriented movement	Movement of PMNs with a directional alignment in a chemotactic response.
Oscillation	Constant reorientation of directional alignment of PMNs.
Polarity	"Head-tail" (monopodial) alignment of PMNs in oriented movement during a chemotactic response.
Pseudopodium	A cytoplasmic extension (projection) arising as a result of cellular deformation and cytoplasmic activity in amoeboid-type motion.
Random movement	A nondirectional movement with no specific polarity and/or orientation.

tion toward or away from chemical substances. The speed with which the PMNs move or the frequency of their turning is referred to as chemokinesis [Keller et al, 1977; Miller and Cheung, 1982; Wilkinson and Allan 1978]. The terminology was coined because of conclusions drawn from visual observations and time-lapsed analysis of PMN movement. With the utilization of videotape and multiple-speed cinemicrographic analyses [Cheung et al, 1982] in conjunction with other cellular probes in our laboratory [Cheung et al, 1983; Kawaoka et al, 1981; Kimura et al, 1981; Miller and Cheung, 1982], a more recent concept on chemotaxis and chemokinesis has been established and will be described in detail in this report. In our interpretation, chemotaxis is defined as the orientation of PMNs to an established chemoattractant gradient and chemokinesis as the enhanced motile activity of PMNs in terms of cellular deformation, oscillation, and translocation. In order to compare our data with previous reports, the relevant terminology is defined in Table I.

Movement of PMNs and the functional characteristics of chemotaxis and chemokinesis are investigated and analyzed in three separate approaches: (1) videotape analysis, (2) elastimetry analysis, and (3) tritonation-reactivation analysis.

VIDEOTAPE ANALYSIS

Introduction

The processes of chemotaxis and chemokinesis require the sensation and translation of information from the environment into a series of complex cellular responses, resulting in directional movement. Recent studies of PMN movement and behavioral characteristics have utilized direct visual (microscopic) observations and/or time-lapse documentation as tools for quantitative investigations on PMN movement [Ramsey, 1972; Zigmond, 1977, 1978a,b; Zigmond et al, 1981]. However, previous studies in our laboratory on protozoan motility demonstrated significant differences between what was perceived visually and what was interpreted with the aid of multiple-speed cinemicrographic and videotape documentations [Brokaw, 1966; Cheung 1973, 1978, 1981; Cheung and Jahn 1975; Cheung et al, 1982; Jahn and Bovee 1969]. Accordingly, we have adapted these microdocumentation techniques to study the movement and behavioral characteristics of PMNs in response to chemoattractants. We have reexamined the phenomenon of PMN movement and the functional characteristics in chemotaxis and chemokinesis with the aid of high-speed and time-lapsed cinemicrography and videotape documentation.

Documentation and Analysis

Cinemicrographs and videotape recordings were made on the movement of PMNs. Nonstimulated PMNs served as random controls (Fig. 1) and PMNs subjected to a chemoattractant gradient as test cells (Fig. 2). The camera (movie or video) was focused at the middle of the bridge of the gradient slide with refocusing as necessary throughout the documentation. The gradient slide was machined from Plexiglas and was significantly modified from Zigmond's (see Appendix for specifications). One drop of PMN suspension (at a concentration of 10^6 cells/ml), which was prepared by a modification from Böyum [1968] and Williams et al [1977], was placed at the center of the bridge, and a coverslip was symmetrically placed on top. One of the two chambers of the slide was carefully filled with PMN suspension medium (McCoy's 5a medium [McCoy's] or Dulbecco's phosphate-buffered saline [PBS]). Care was taken to establish a continuum between the fluid on the bridge and the fluid in the filled chamber, without causing overflow into the adjacent empty chamber. The gradient slide was then placed on the microscope stage and incubated at 37°C by a Sage air curtain for 15 minutes. Chemoattractant (10^{-8} M N-formyl-methionylleucylphenylalanine [F-Met-Leu-Phe]) was then carefully added to the empty chamber, resulting in the establishment of a concentration gradient [Cheung et al, 1982]. The control preparation consisted of PMNs on the bridge and both chambers were filled with the

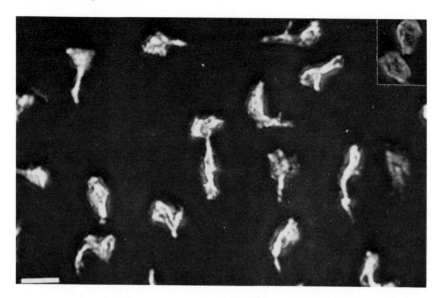

Fig. 1. Random PMN movement. The PMNs were suspended in McCoy's culture medium and incubated at 37°C. Note the random directional orientation and the extensive cellular deformation. Corner insert shows two PMNs suspended in an identical medium at room temperature. These two cells were not deformed and both retained the spherical inactive configuration [Cheung et al, 1982]. Magnification, ×645; optics, phase contrast; reference bar, 10 μm.

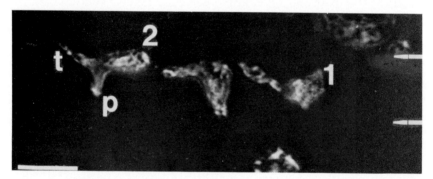

Fig. 2. Chemotactic response. Three PMNs responding to an established 10^{-8} M F-Met-Leu-Phe gradient. The cells were all moving toward the gradient and a cytoplasmic trail was detached and left behind (as indicated at t). Pseudopodium 1 was moving toward the gradient and the chemotactic response of this cell was scored as positive. Pseudopodium 2 was moving toward the gradient, while pseudopodium p had just initiated its sideways movement. Since the overall directional movement of the cell was toward the gradient, the chemotactic response was still scored as (positive). Note the bipodial and nonpolarized configuration of the cell with pseudopodia 2 and p as opposed to the monopodial and polarized configuration of the first cell [Cheung et al, 1982]. Magnification, ×645; optics, phase contrast; reference bar, 10 μm.

PMN suspension medium (McCoy's or PBS). Movement and responses of the PMNs were documented and later analyzed for quantitative and time-dependent behavioral characteristics.

Videotapes/cinemicrographs were analyzed on a monitor/projection screen. Pseudopodial formation, cytoplasmic movement, direction of cellular movement, and the actual path traveled by the PMNs were traced. PMNs orienting and moving toward the chemoattractant gradient were scored as positive responses (+) and cells orienting and moving away were scored as negative (−). PMNs which were not oriented or moving in either direction were scored as neutral or zero (0) response (Fig. 2).

Usually, pseudopodia would extend in the same direction of overall movement of a PMN. Rarely, however, a PMN demonstrated overall movement toward the chemoattractant gradient, but simultaneously might be extending one or more pseudopodia in other directions (Fig. 2). In such cases, the direction of overall movement took priority in the scoring.

The net fractional response at a specific time was then determined as the difference between the positive (+) and negative (−) responses. This could be net (positive) or (negative), depending on the orientational behavior of the PMNs to the chemoattractant. A plot of this net fractional response with time represents the standard time-dependent response curve of normal PMNs to the chemoattractant.

Results

PMNs, placed on the bridge of our gradient slide at room temperature (in McCoy's or PBS), were initially round, nondeformed, and nonlocomotory. Deformation started after 5–15 minutes' incubation at 37°C and was normally followed by random movement. Figure 1 shows the typical random movement of incubated but unstimulated (ie, without stimulation by a chemoattractant) PMNs. Forward movement usually began with the formation of a single pseudopodium and streaming of cytoplasmic granules. Occasionally, multiple pseudopodia might start out. The direction of pseudopodial formation was not limited to the front end of the cells. Pseudopodia might also arise by the sides—changing a cell from a simple monopodial state (the so-called head-tail configuration) to a bi- or tripodial (polypodial) state (Figs. 1–3). Each pseudopodium could be retracted at any time and a pseudopodium might even arise at the posterior portion of any PMN. As can be seen in Figure 1, most of the moving cells were not aligned in the "head-tail" configuration or in a polarized state. In the unstimulated condition (ie, in the absence of a chemoattractant gradient), the direction of locomotion of the PMNs was completely inconsistent; a movement pattern normally referred to as

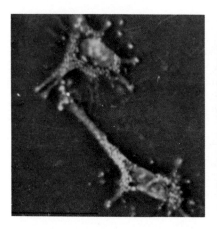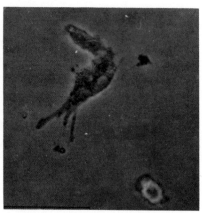

Fig. 3. Active deforming PMNs. The three PMNs were actively deforming and had attained a polypodial state, complete with a loss of orientation polarity. Note the multiple trail filaments arising as a result of rapid cell movement and active retraction of pseudopodia [Cheung et al, 1982]. Magnification, ×1,000; optics, phase contrast; reference bar, 10 μm.

random. Each cell could move in any direction at any given time. Each advancing pseudopodium could retract and reform, partially and/or completely, apparently at will. Scoring of typical unstimulated and nonoriented random movement is shown in Figure 4 (control response curve).

Introduction of the chemoattractant F-Met-Leu-Phe (at 10^{-8} M concentration) into the empty chamber of the gradient slide (see Documentation and Analysis) established a concentration gradient and initiated a chemotactic response. The first noticeable change was the sudden enhancement of cytoplasmic streaming and membrane deformation. The cells seemed to have been instantaneously activated by the presence of the chemoattractant. Initial enhancement of cytoplasmic and cellular activities was brought about by the 37°C incubation. However, on a time-dependent scale, the activated cellular movement and cellular activities brought about by the chemoattractant were accelerated and were of a greater magnitude in comparison with PMNs in random movement.

The overall behavioral response of the PMNs to F-Met-Leu-Phe was both chemotactic (enhanced oriented directional movement) and chemo-kinetic (enhanced cellular and locomotory activity). The response of PMNs 5 minutes after exposure to a chemoattractant gradient is shown in Figure 2, and a typical chemotactic response curve is shown in Figure 5.

During a typical chemotactic response, there was rarely a consistent "head-tail" appearance. Apparent polarity of the "head-tail" configura-

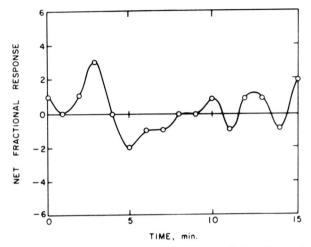

Fig. 4. Random control curve of PMNs. The curve was obtained by scoring the directional orientation and movement of 14 PMNs under investigation. The 14 cells were shown in Figure 1 [Cheung et al, 1982].

tion was observed less than 10% of the time. When the cells were moving quickly, more than one posterior end ("tail") could be formed (Fig. 3), but they could be retracted instantaneously. At times, the forward movement could be so fast (more than one cell-length per minute) that the adhering posterior end (or ends) and retracting pseudopodium (or pseudopodia) could not catch up with the fast locomotion. Consequently, long, thin filamentous retraction fibers would be detached and left behind. Such a detachment process with pseudopodial retraction is shown in Figure 6. It is also observed that during a typical response, a lack of constant orientation and "head-tail" polarity was observed. The overall movement of cells in a given field was always toward the direction of the approaching chemoattractant gradient, but any individual cell failed to show a sustained directional orientation or a sustained directional movement toward the gradient. Most, if not all, of the cells started to move in the direction of the chemoattractant. They soon, however, began to move sideways, forward-and-backward, or in the reverse direction. Only very rarely did they simply continue to move forward undisturbed. Normally, within 1 minute, a cell that had moved away from its original direction usually reoriented and moved again in the direction of the approaching chemoattractant. The overall movement was, therefore, toward the direction of the gradient. During a typical chemotactic and chemokinetic response, PMNs which had moved away from the gradient tend to reorient back to the direction of the approaching gradient. This type of reorientation was oscillatory in

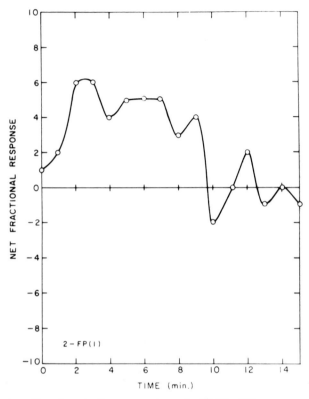

Fig. 5. A positive chemotactic response curve for PMNs. This response curve represents the behavioral (chemotactic) response of 11 PMNs to a F-Met-Leu-Phe (10^{-8} M) gradient, scored at 1-minute intervals.

nature and could be revealed clearly in actual cell path tracings (Fig. 7). Such movement characteristics were also shown in a time-dependent analysis of the chemotactic response curve. A comparison of the curves obtained from 1-minute tracing analysis of the chemotactic activity (as shown in Fig. 5) with 1/4- and 1/2-minute scorings revealed these unexpected time-dependent oscillatory features which were not apparent in visual scoring or time-lapsed analysis. In chemokinesis, enhanced frequency of oscillation and amplitude of net fractional response were significant characteristics typically found in a time-dependent fractional response curve. Such characteristics are clearly illustrated in Figure 8.

Discussion

Motile behavioral responses of PMNs have been classified into two categories—chemotaxis and chemokinesis. Chemotaxis refers to the orien-

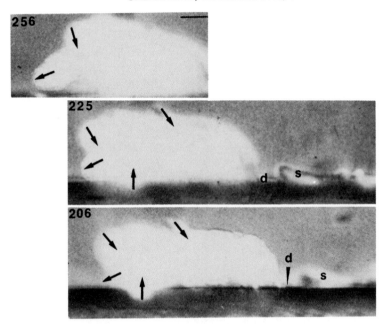

Fig. 6. Cross-sectional view of PMN movement. These three frames (frames #206, 225, and 256) are taken from a 16-mm movie filmed at 16 frames per second. The frames are cropped and staggered to show their relative position in the locomotion. The arrows indicate the direction of pseudopodial formation/retraction. During the formation of one forward pseudopodium, side and top pseudopodia are being retracted. Note the detachment of a trailing cytoplasmic segment (retraction filament) during rapid locomotion (d indicates the point of detachment and s the segment detached). The amoeboid movement exhibited by a PMN (cross-sectional view) differs a bit from the types exhibited by *Chaos chaos* and *Hyalodiscus simplex;* but is very similar to the movement exhibited by *Amoeba proteus* [unpublished data]. Magnification, ×1,000; optics; phase contrast; reference bar, 1 μm.

tation of the cells to the established and maintained gradient of a chemoattractant. This is exhibited in vivo as accumulation of cells at the site of infection or injury, and in vitro as cellular orientation and alignment toward the chemoattractant gradient. Chemokinesis refers to the motile activity of the cells in terms of translocation. This is exhibited in vivo as increased rate of movement toward the stimulus and in vitro as enhanced intracellular cytoplasmic activity, orientational oscillation, and locomotion.

Most visual assays reported in the study of PMN movement have utilized time-lapsed photography and/or direct visual scoring of PMNs in motion. Visual scoring is subjective and difficult to carry out even under ideal conditions. These difficulties are compounded when the observer has

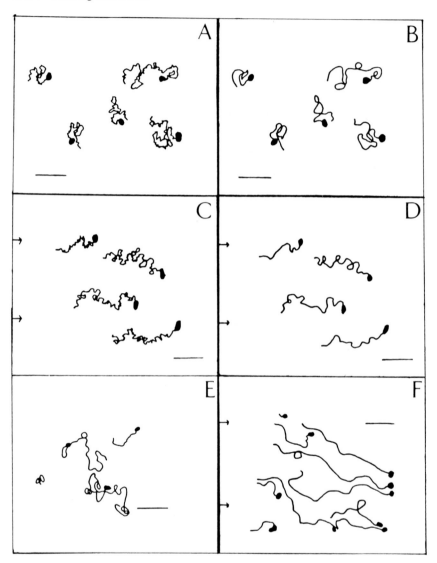

Fig. 7. Path tracings of PMNs in motion (the cells indicate the starting points of tracings). A. Tracings of the path of five cells (PMNs) in the absence of chemoattractant. The tracings were made at 5-second intervals. Note the random motility exhibited. Reference bar, 100 μm. B. Identical tracings of A with the tracings made at 30-second intervals. Reorientations which can be seen easily in A are not apparent with such a big time lag; however, the actual directional path of movement is still identical with that in A. C. Tracings of the path of four cells (PMNs) during chemotaxis. The tracings were made at 5-second intervals. The arrows indicate the direction of the approaching chemoattractant

to evaluate many cells (20 or more) at a time, each with time-dependent variations in individual motions.

The videotape documentation technique which we used for movement investigations solved most optical and visual problems in time-dependent motility studies and still provided high-quality records for motility analysis. Behavioral movements of PMNs were recorded for later analysis and evaluation. Double-blind analysis ensured objective scoring of time-dependent events. Also, videotapes provided permanent records for future reference or further study and retracing.

The establishment of a chemoattractant gradient is a critical prerequisite for the in vitro study of PMN movement. In the absence of such a gradient (as in the case when PMNs are put into a chemoattractant solution instead of being placed in the approaching path of a chemoattractant gradient), an oriented chemotactic response will not take place, and the cells will only exhibit chemokinesis. Figure 9 illustrates such a phenomenon. The gradient slide (developed in our laboratory for these studies) provided a dependable concentration gradient and was designed to permit PMNs to detect the lowest possible concentration of chemoattractant [Cheung et al, 1982].

Previous reports of direct observation and time-lapsed study of PMNs during chemotaxis suggested relatively sustained orientation of cells and an apparent polarity with a "head-tail" configuration. It was also reported that, when compared with randomly moving cells, paths followed by such

---

gradient. Note the increased frequency of reorientations and the overall directional path. Reference bar, 100 μm. D. Identical tracings of C with the tracings made at 30-second intervals. Again, reorientations are not apparent with such a big time lag; however, the actual directional path of movement is still identical with that in C. The time-dependent reorientations and oscillatory patterns in chemotaxis cannot be picked up in the 30-second time lag. E. Ramsey's tracings (from time-lapse movies) of the paths of some cells (PMNs) in the absence of chemoattractant (modified from Ramsey [1972]). Reference bar, 100 μm. These paths differ from the oscillatory paths of cells traced at 5-second intervals (as shown in A), but are remarkably similar to paths of the same cells traced at 30-second intervals (as shown in B). F. Ramsey's tracing (from time-lapse movies) of the paths of some cells (PMNs) during chemotaxis. The arrows indicate the direction of the approaching chemoattractant gradient (modified from Ramsey [1972]). Reference bar, 100 μm. Again, these paths differ from the oscillatory paths of cells (during a chemotactic response) traced at 5-second intervals (as shown in C), but are remarkably similar to paths of the same cells traced at 30-second intervals (as shown in D). Comparison of A, B, C, and D with E and F confirms the similarity of the paths of movement. The difference between the actual shape of the paths (oscillations and reorientations in A and C and the apparent absence of oscillations and reorientations in B, D, E, and F) lies in the fact that the tracings were made at different time intervals (5 seconds for A and C and 30 seconds for B, D, E, and F). The time-dependent reorientations and oscillatory patterns in cellular movement and chemotaxis cannot be picked up in the 30-second time lag [Cheung et al, 1982].

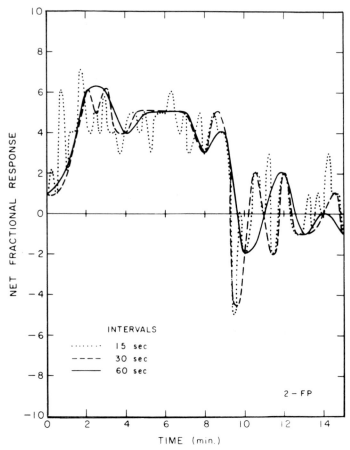

Fig. 8.   Time-dependent chemotactic response curves for PMNs. The solid curve is the
same curve in Figure 5 and represents the response of PMNs to a 10^{-8} M F-Met-Leu-Phe
gradient, scored at 1-minute intervals. Response curves of 1/4- and 1/2-minute interval
scorings of the solid curve are superimposed to show orientational and movement changes
(oscillations) during the chemotactic response.

oriented cells revealed decreased frequency and angles of turning, ie,
decreased chemokinesis [Ramsey, 1972; Wilkinson and Allan, 1978;
Zigmond, 1977, 1978a, b; Zigmond et al, 1981].

Our results differ significantly from their studies. When analyzed by
multiple-speed cinemicrographic and videotape techniques, constant reorientation (oscillation) of individual cells occurred (Fig. 8). Polarity of
any individual cell was inconsistent and only rarely occurred (Figs. 1,3).
Frame-by-frame tracing of paths followed by individual cells moving
toward a chemoattractant showed that chemokinesis was actually

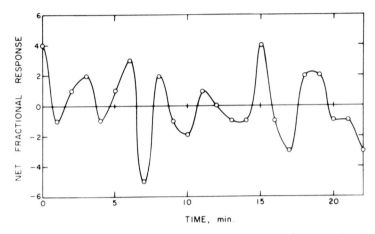

Fig. 9. Chemokinetic curve for PMNs in F-Met-Leu-Phe (10^{-8} M) solution. A typical behavioral response (chemokinetic, but not chemotactic) of PMNs suspended in a chemoattractant solution without gradient establishment. Note the random orientation as opposed to the chemotactic response curves of Figures 5 and 8. The curve is similar to Figure 4 (random control curve) for directional orientation (ie, absence of chemotaxis); however, the enhanced amplitude of the fractional response is characteristic of chemokinesis [Cheung et al, 1982].

increased (Fig. 7). This was consistent with the increased (enhanced) amplitude and frequency of oscillations shown in the scoring curve (Fig. 8). When the paths of cells were traced and their orientation scored at 1-minute intervals (instead of 1/12-, 1/4- or 1/2-minute intervals we normally used), the paths and the chemotactic response curves would be identical to those described by other investigators [Cheung et al, 1982]. By analyzing the data at much shorter time intervals, therefore, a more detailed and accurate sequence of time-dependent events occurring during the chemotactic response can be obtained.

We have shown that analysis of time-dependent characteristics of the chemotactic response is essential to the full understanding and characterization of the chemotactic, chemokinetic, and movement activities of PMNs. This videotape analytical technique can provide a sensitive probe with which to dissect the various perturbations of PMN movement in man.

ELASTIMETRY ANALYSIS

Introduction

A number of cellular events have been determined to associate with PMN movement—among other cellular activities, alterations in the shape

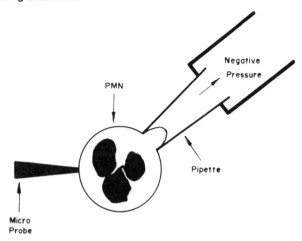

Fig. 10. Schematic drawing of the technique of cell elastimetry.

of the cell and the functional state of the membrane are considered prerequisites for movement. A functional deformation takes place in the membrane during movement; when placed in a chemoattractant gradient, PMNs are activated to deform and move. Along with other cellular events in chemotaxis and chemokinesis, our interest has focused upon deformability and its relationship to PMN movement.

Previous studies of membrane alteration in motile unicellular organisms (such as amoebae) have shown a possible correlation between membrane deformability and amoeboid movement [Jahn and Bovee, 1969; Miller and Cheung, 1982]. Lichtman [1970] studied deformability of human bone marrow granulocytes by the technique of cell elastimetry and revealed that increasing deformability correlated with granulocyte egress from marrow.

The technique of cell elastimetry measures the amount of negative pressure required to aspirate a cell into a micropipette. The basic principle is illustrated in Figure 10. The higher the negative pressure required to aspirate the cell, the more resistant the cell is to deformation. A highly deformable cell requires relatively little negative pressure to aspirate it into a micropipette.

Cell Elastimetry

Cell elastimetry is a single cell assay and is a technique best equipped to study the effects of chemotactic factors (chemoattractants) on cell motility. We have adapted this method to the study of the deformability and movement of human peripheral blood PMNs.

Cell elastimetry was performed (with minor modifications) as previously described [Kawaoka et al, 1981; Lichtman, 1970]. Briefly, micropipettes of 3–5-μm internal diameter were drawn from glass capillary tubing and mounted in a holder (Fig. 10). PMNs (obtained as described earlier) were incubated for 15 minutes at 1×10^6 cells/ml concentration in McCoy's (plain McCoy's as control and McCoy's containing varying concentrations of F-Met-Leu-Phe for chemoattractant trials). Several drops of the PMN suspension were placed in a Leitz moist chamber and diluted with 10 drops of McCoy's (containing the same concentration of chemoattractant). The micropipette was connected to a source of negative pressure, and a PMN was positioned at the tip with the assistance of a Leitz micromanipulator. Slowly increasing negative pressure was applied and the negative pressure required to completely aspirate the PMN into the micropipette was then noted.

For deformation studies involving chemoattractants, the deformability of the PMNs was measured after 15–30 minutes' incubation in the chemoattractant-containing medium.

Results

For the elastimetry studies, deformability was measured and expressed as centimeters of mercury negative pressure (cm Hg) to completely aspirate a PMN into a micropipette. For this series of experimentation, the mean negative pressure for normal PMNs was 11.5 (± 3.2 SD) cm Hg. This result is in close agreement with the mean of control populations of cells from other studies in our laboratory. At a chemoattractant (F-Met-Leu-Phe) concentration as low as 10^{-11} M, a change in the deformability characteristics of a population of PMNs was demonstrated. The increase in deformability became more marked at peptide concentrations in the range of $1–5 \times 10^{-9}$ M, corresponding to the optimal concentration range in the filter (chemotaxis) assay. Concentrations at 10^{-8} M and greater were associated with significant increases in the negative pressure required to aspirate the cells. The concentration dependence of the deformability of PMNs on F-Met-Leu-Phe has been described in detail in a previous report [Kawaoka et al, 1981] and is shown in Figure 11.

Discussion

Studies on membrane deformation and cell motility in amoebae (*Chaos chaos, Amoeba proteus*, and *Hyalodiscus simplex*) have shown the role of deformability in cellular movement. The large amoeba, *Chaos chaos*, is efficient in movement and its membrane is extremely deformable. The smaller amoebae, *Amoeba proteus* and *Hyalodiscus simplex*, are not as

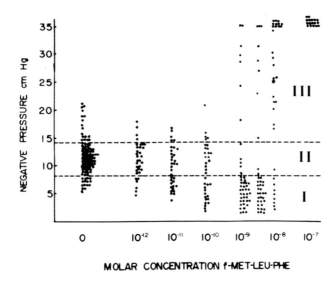

Fig. 11. Concentration dependence of the deformability of PMNs incubated in F-Met-Leu-Phe [Kawaoka et al, 1981].

efficient in movement (locomotion) and their membranes are less deformable. Decrease in deformability results in decrease in pseudopodial formation, thus putting the motile cell at a mechanical disadvantage in movement maneuvers.

Cell elastimetry studies on PMNs suggest a correlation between deformability and cell motility but do not establish the primary effect of chemotactic stimuli upon deformability. In other words, the elastimetry assay might simply be measuring a result rather than a central event of the membrane process. We have probed this question in two different ways.

First, the role of deformability has been studied with the aid of our videotape and cinemicrographic assay, as described earlier [Cheung et al, 1982]. Tracing analysis shows that a highly deformable membrane increases the facility of a PMN to move. Analysis of deformability and movement with reference to varying concentrations of chemoattractants confirms the increased deformability of the PMN membrane in a dose-dependent fashion and that movement of PMNs is closely related to a chemotactic factor-induced increase in membrane deformability. Second, the deformability and movement of neonatal PMNs were studied. Neonatal PMNs were not effective in function and were not chemotactic-responsive. Neonatal PMNs were found to be significantly less efficient in movement. It was also found that there was a marked decrease in deformability. It can now be concluded, therefore, that deformability is a prerequisite to PMN movement and that decreased (impaired) deformability (as demonstrated in neonatal PMNs) results in compromised cellular movement.

A highly deformable membrane increases the facility of a cell to move. Chemoattractants induce the deformability of the cell membrane (activation) in a dose-dependent fashion, and the movement of PMNs is closely related to a chemotactic factor-induced increase in membrane deformability.

By utilizing the technique of cell elastimetry, we have shown a correlation between chemoattractant (F-Met-Leu-Phe) concentration and PMN deformability which parallels the effects of the chemoattractant on filter (chemotaxis) assay [Kawaoka et al, 1981]. The experimental results indicate that alterations in PMN deformability occur as a natural consequence of stimulation of human PMNs with a chemotactic factor and that deformability is a prerequisite to efficient movement. Abnormalities in deformability can lead to impaired PMN movement and related problems in host defense.

TRITONATION-REACTIVATION ANALYSIS
Introduction

The mechanisms by which PMNs migrate to inflammatory sites (chemotaxis, chemokinesis, and movement) are incompletely defined. The mechanisms require the sensation and translation of information from the environment into a series of complex responses, resulting in directional movement. A major question involves the sequence by which specific extracellular stimuli induce directed movement of PMNs. Of particular importance in characterizing this response is the determination of the degrees to which the movement and orientation mechanisms are self-contained within the cell and the degrees to which they are under membrane control. These studies were suggested by previous work on reactivated models in cilia and flagella [Cheung, 1975, 1977; Goldstein, 1974; Naitoh and Kanebo, 1972, 1973] and the movement reactivation work on cellular contraction and cytokinesis of Burnside, Cande, and Hoffman-Berling [Burnside et al, 1980, 1982; Cande, 1980; Hoffman-Berling, 1954a, b].

Tritonation-Reactivation

The basic technique of tritonation-reactivation was one of partial destruction of the cell membrane by a detergent (partial demembranation), thus providing small openings in the membrane through which ATP and necessary ions could achieve unrestricted passage. The presence of

ATP served as the energy source to power the still-intact (and functional) internal movement mechanism of the PMNs. The success of the procedure depends on the tritonation (demembranation) process; tritonation has to be carried out to an extent where the partial demembranation is extensive enough to provide adequate passage for ATP and the necessary ions, but not enough to cause drastic membrane destruction (eg, cytolysis) or dissociation of the cytoskeletal structure responsible for movement.

This approach was adopted from earlier studies in which utilization of a nonionic detergent, octylphenoxyl-polyethoxyethanol (commercially known as Triton X-100), on ciliates provided entire "pure ciliate models" capable of being reactivated to locomote by the introduction of ATP [Cheung, 1977; Naitoh and Kanebo, 1972, 1973].

To fully understand the fundamental kinetics of PMN movement and behavioral response to chemoattractants, it would be crucial to develop a "pure PMN model" system which could be subjected to experimentation. Such a "pure PMN model" must be capable of generating motility data as close to live (movement) activities of the PMNs as possible, but, at the same time, could serve to separate or eliminate the multifaceted effects of the movement- and orientation-contributing factors and related complications.

Procedural Details

PMNs were obtained and separated from venous blood as previously described [Böyum, 1968; Williams et al, 1977]. However, hypotonic shock was not carried out to ensure integrity of the PMN membrane. A final concentration of a yield of $1–5 \times 10^6$ PMNs/ml McCoy's or PBS was obtained.

Two drops of PMN suspension were put onto the bridge of a gradient slide [Cheung et al, 1982]. A coverslip (#1 or 1–1/2) was placed symmetrically on top of it. PBS (or suspension medium used) was added to each chamber of the slide (on both sides of the bridge), and the PMNs were incubated and maintained on the stage at 37°C by a Sage air curtain until cellular deformation and locomotion were observed. After cellular deformation and movement were confirmed, the suspension medium was then carefully replaced with the tritonation medium (0.5% v/v Triton X-100). Because of the design and the geometry of the gradient slide, fluid removal and introduction could be performed with ease. The tritonation process was maintained for 15 seconds. The tritonation medium was carefully removed, and the suspension medium was gently flushed over the cells immediately after removal of the tritonation medium [Stich et al, 1981]. The flushing/washing procedure was repeated several times to ensure complete removal of the detergent.

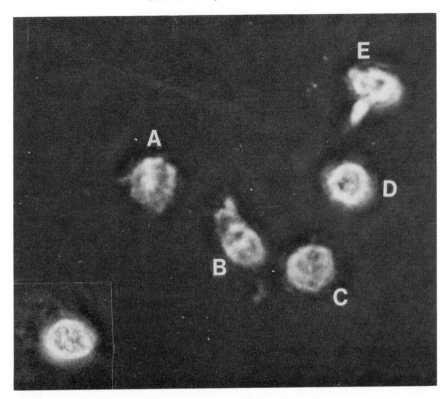

Fig. 12. Photomicrograph (35 mm) of 6 PMN models taken 2 minutes after ATP reactivation. Five of the models are labeled (A–E) for later movement reference. Note the deformation and pseudopodial formation in some of the cells and the actual cellular locations. Magnification, ×400 (photographically enhanced to give a magnification of about ×1,000); optics, phase contrast. Corner insert (not labeled) is the unreactivated model control.

Four identical tritonated preparations were made before use for each experimental run; two were kept as controls to ensure inability of the tritonated models to move, while the others were reactivated by the introduction of the reactivation medium (5 mM ATP, 4 mM Mg^{++}, 0.1 mM Ca^{++}). The entire reactivation procedure was observed under the microscope and was simultaneously videotaped for later analysis. Black-and-white 35-mm still photomicrographs were taken at various times for reference and confirmation (Figs. 12–14).

Reactivated PMNs (with tritonated, normal, and normal/chemotactic controls) were subjected to the influence of an established chemoattractant gradient, set up by 10^{-8} M F-met-Leu-Phe [Cheung et al, 1982]. The

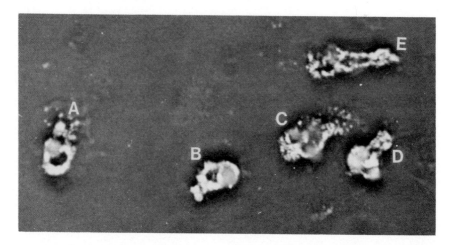

Fig. 13. Photomicrograph of the five reactivated models taken 5 minutes after ATP reactivation and 1 minute after the introduction of a chemoattractant (10^{-8} M F-Met-Leu-Phe) from the right side. Reactivated cellular movement is extensive (note the movement of the cells with reference to position-location in Fig. 12), but a chemotactic (directional) response is absent. Magnification and optics same as in Figure 12.

Fig. 14. Photomicrograph of same models taken 15 minutes after reactivation (which is 11 minutes after being subjected to a chemoattractant gradient). Reactivated movement is excellent, but a chemotactic response is absent. Magnification and optics same as in Figure 12.

behavioral response of the reactivated PMNs to the chemoattractant gradient was again videotaped for later analysis.

Results

PMNs circulating in the blood stream or placed on a glass slide in standard culture medium (without incubation) were usually round, nondeformed, and nonlocomotory, exhibiting their normal spherical configuration. After sustained air curtain (37°C) incubation, the cells began to deform. Deformation was an indication of cell motility and was characterized by cytoplasmic streaming and pseudopodial formation.

Upon contact with Triton X-100, the pseudopodia retracted and the deformed PMNs immediately resumed their inactive spherical configuration (see Fig. 12 insert). The tritonated cells (referred to in the literature as "models") did not move, deform or show any sign of cytoplasmic or membrane activity, even after prolonged periods of 37°C air curtain incubation (control). Following tritonation, the models were confirmed to be nonmotile and nonfunctional.

Within 1–2 minutes after the introduction of the reactivation medium, the cells began to deform (Fig. 12). Successful reactivation of the PMN models was characterized by reestablishment of cytoplasmic activity, pseudopodial formation, and cellular deformation. Cellular translocation (movement with locomotion) normally occurred in successful reactivations, while partial reactivations resulted only in limited movement and partial pseudopodial formation. Reactivated activities peaked in 4–5 minutes and normally could last for 10–15 minutes (Figs. 13, 14). The percentage of success, the degree of reactivation, and procedural details of the tritonation-reactivation technique have been described and discussed in previous reports [Cheung et al, 1983; Stich et al, 1981].

Discussion

The establishment of this tritonation-reactivation technique provides a probe with which to investigate the functional mechanics of chemotaxis and chemokinesis and also serves to supply a tool to dissect and analyze disorders of PMN movement [Cheung et al, 1983].

We selected two disorders of human PMN movement (believed to result from defects at two different levels of the cell) to interpret with normal PMN model reactivation results. The two disorders of human PMN movement chosen were human cord blood PMN (cord-PMN) and PMNs from Chediak-Higashi patients (CH-PMN). Cord-PMNs have previously been demonstrated to have multiple defects best related to membrane dysfunction. The abnormalities include movement and, in addition, decreased deformability, impaired redistribution of surface adhesion sites, impaired lectin- and C5a-induced aggregation, and impaired concanavalin A (Con A) induced capping [Miller, 1975, 1978; Kimura et al, 1981]. CH-PMNs are believed to have movement defects resulting from impaired microtubule (cytoskeletal) function. In contrast to cord-PMNs, CH-PMNs show spontaneous Con A-induced capping [Boxer et al, 1976].

Normal adult PMNs demonstrate random movement and pseudopodial formation and are chemotactic-responsive to F-Met-Leu-Phe (or other chemoattractants). After tritonation, the models are nonfunctional; movement as well as pseudopodial characteristics are lost. The reactivation process helps to restore the movement and pseudopodial activities. However, the reactivated models do not regain their chemotactic responsiveness. The tritonation process leaves the cytoskeletal structures (intrinsic movement mechanism) functionally intact, but the same process partially and selectively destroys the cell membrane—enough to eradicate the chemotactic sensation/integration function of the membrane.

Cord-PMNs also demonstrate random movement and pseudopodial formation under physiological conditions, but they are normally not chemotactically responsive. Like normal adult PMNs, tritonated cord-PMNs lose their movement and pseudopodial characteristics during tritonation and regain them during ATP reactivation. This comparison suggests that the intrinsic or cytoskeletal determinance of movement in cord-PMNs is functionally as intact and mature as in adult PMNs; their chemotactic (movement) defect lies directly or indirectly at the membrane level.

In contrast to cord-PMNs, CH-PMNs do not exhibit random movement or pseudopodial formation under normal conditions, nor are they chemotactically responsive. Their tritonated models cannot be reactivated to move or translocate. The cytoskeletal structure (movement mechanism) of CH-PMNs is simply not functional.

The tritonation-reactivation technique is an excellent cellular probe which can serve to generate "pure models" for experimentation and to be used to dissect and analyze the various functional impairments (at various cellular levels) of PMN movement.

We have investigated the ATP-reactivated movement of partially demembranated PMNs (tritonated models) and find that their basic movement mechanism is a physical and self-contained component (the cytoskeletal structure), while the orientation mechanism which is responsible for a full directional (chemotactic) response is under membrane control. With the utilization of this technique, the levels of chemotactic and movement dysfunctions of some PMNs (cord and CH in this report) have been successfully characterized.

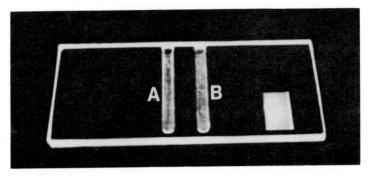

Fig. A.1.   The gradient slide (see Appendix for specifications).

CONCLUSIONS

The above-described analytical techniques (videotape, elastimetry, and tritonation-reactivation) are excellent probes which can be utilized to investigate and analyze PMN movement. A combination of these techniques with other cellular probes (eg, capping, phagocytosis) can serve to provide an extensive survey on the functional competence of PMNs.

ACKNOWLEDGMENTS

The research was supported in part by the National Institutes of Health, National Institute of Child Health and Human Development, Program Project 1-P01-HD10975-01A1, and the National Institutes of Health grant 5-T32-A1-07014. The research was also sponsored by a grant from the Dean's Office of the University of California, Davis School of Medicine, and a Faculty Research support grant from the University of California at Davis.

APPENDIX: THE GRADIENT SLIDE

A slide, significantly modified from Zigmond [Cheung et al, 1982; Zigmond 1978a,b] was machined from Plexiglas. A typical slide was made from 2-mm Plexiglas (can be thinner, if desired) and machined to satisfy the following general specifications: (1) the two chambers (A and B) had to be of identical dimensions (see Fig. A.1); (2) each chamber had to be at least 3.5 mm in width and over 35 mm in length, with a length-width ratio of at least 10:1; (3) the volume of each chamber had to be at least 125 μl in capacity; (4) the edges of the chamber had to be evenly buffed; (5) the bridge between the two chambers had to be of a width of at least 7 mm and the length-width ratio to be over 5:1 (normally about 6:1); and (6) the coverslip used had to be of such a size and be placed in such a manner that the four rounded corners of the chambers were symmetrically exposed.

The general limiting conditions were defined theoretically, taking into account surface tension, wall effects, molecular size, coefficient of molecular diffusion of particulates, low Reynold's number fluid mechanics, and gradient sensitivity. The theoretical predictions were verified by diffusion studies (tracer and fluorescence manipulations) and confirmed by trial runs before experimental use (see Cheung et al [1982], for theoretical details).

REFERENCES

Boxer LA, Watanabe AM, Rister M, Besch HR, Allen M (1976): Correction of leukocyte function in Chediak-Higashi syndrome. N Engl J Med 295:1041–1053.

Böyum A (1968): Isolation of mononuclear cells and granulocytes from human blood. Scand J Clin Lab Invest 21(Suppl 97):77–89.

Brokaw CJ (1966): Mechanisms and energetics of cilia. Am Rev Respir Dis 93:32–40.

Burnside B, Smith B, Nagata M (1980): Reactivation of contraction in detergent-lysed teleost retinal cones. J Cell Biol 87(2 Pt 2):200a.

Burnside B, Smith B, Nagata M, Porrello K (1982): Reactivation of contraction in detergent-lysed teleost retinal cones. J Cell Biol 92:199–206.

Cande WZ (1980): A permeabilized cell model for studying cytokinesis using mammalian tissue culture cells. J Cell Biol 87:326–335.

Cheung ATW (1973): Determination of ciliary beat in *Opalina obtrigonoidea* and in rabbit tracheal explants. Doctroal Thesis (UCLA), University Microfilm, Ann Arbor, Michigan.

Cheung ATW (1975): Reactivation of glycerinated cilia from *Opalina*. Acta Protozool 14:99–104.

Cheung ATW (1977): Reactivation of tritonated models of *Opalina*. Acta Protozool 16:377–384.

Cheung ATW (1978): Ciliary activity of stationary *Opalina*. Acta Protozool 17:153–162.

Cheung ATW (1981): High speed cinemicrographic and microvideo analyses of oviductal ciliary activity. Wasmann J Biol 39:68–78.

Cheung ATW, Jahn TL (1975): Helical nature of the continuous ciliary beat of *Opalina*. Acta Protozool 14:219–232.

Cheung ATW, Miller ME, Keller SR (1982): Movement of human polymorphonuclear leukocytes: A videotape analysis. J Reticuloendothel Soc 31:193–205.

Cheung ATW, Miller ME, Kimura GM, Kawaoka EJ, Keller SR (1983): Movement reactivation of tritonated models of human neutrophils. Fed Proc (in press).

Clark RA (1978): Disorders of granulocyte chemotaxis. In Gallin JI, Quie PQ (eds): "Leukocyte Chemotaxis: Methods, Physiology and Clinical Implications." New York: Raven Press, pp 129–356.

Goldstein SF (1974): Isolated, reactivated and laser-irridated cilia and flagella. In Sleigh MA (ed): "Cilia and Flagella." New York: Academic Press, pp 111–130.

Hoffman-Berling H (1954a): Adenosinetriphosphate als betriebsstoff von zellbewegungen. Biochim Biophys Acta 14:182–195.

Hoffman-Berling H (1954b): Die Glycerin-wasserextrahierte teleophaezelle als modell der zytokinese. Biochim Biophys Acta 15:332–339.

Jahn TL, Bovee EC (1969): Protoplasmic movement within cells. Physiol Rev 49:793–862.

Kawaoka EJ, Miller ME, Cheung ATW (1981): Chemotactic factor-induced effects upon deformability of human polymorphonuclear leukocytes. J Clin Immunol 1:41–44.

Keller HU, Wilkinson M, Abercrombie PC, Becker EL, Hirsch JC, Miller ME, Ramsey WS, Zigmond SH (1977): A proposal for the definition of terms related to locomotion of leukocytes and other cells. Cell Biol Int Rep 1:59–64.

Kimura GM, Miller ME, Leake RD, Raghunathan R, Cheung ATW (1981): Reduced concanavalin A capping of neonatal polymorphonuclear leukocytes (PMNs). Pediatr Res 15:1271–1273.

Lichtman MA (1970): Cellular deformation during maturation of the myeloblasts—Possible role in marrow egress. N Engl J Med 283:943–948.

Mease AD, Fischer GW, Hunter KW, Ruymann FB (1980): Decreased phytohemagglutin-in-induced aggregation and C5a-induced chemotaxis of human newborn neutrophils. Pediatr Res 14:142–416.

Metchnikoff E (1893): Lectures on the Comparative Pathology of Inflamation. London: Kegan, Paul, Trench, Truber and Company.

Miller ME (1975): Pathology of chemotaxis and random mobility. Semin Hematol 12:59–82.

Miller ME (1978): Host Defense of the Human Neonate. New York: Grune and Stratton.

Miller ME, Stiehm ER (1981): Infectious Disease of the Fetus and Newborn Infants. New York: Grune and Stratton.

Miller ME, Cheung ATW (1982): Functional alterations in the membrane of motile polymorphonuclear leukocytes. Am J Pediatr Hematol/Oncol 4:77–82.

Naitoh Y, Kanebo H (1972): Reactivated triton-extracted models of *Paramecium canda-tum:* Modification of ciliary movement by calcium ions. Science 176:523–524.

Naitoh Y, Kanebo H (1973): Control of ciliary activities by adenosinetriphosphate and divalent cations in triton-extracted models of *Paramecium candatum.* J Exp Biol 58:657–676.

Ramsey WS (1972): Analysis of individual leukocyte behavior during chemotaxis. Exp Cell Res 70:129.

Synderman R, Pike MC (1977): Disorders of leukocyte chemotaxis. Pediatr Clin North Am 24:377–393.

Stitch MA, Cheung ATW, Miller ME (1981): ATP reactivation of tritonated models of human polymorphonuclear leukocytes. Wasmann J Biol 39:50–55.

Wilkinson PC, Allan RB (1978): Assay systems for measuring leukocyte locomotion. In Gallin JI, Quie P (eds): "Leukocyte Chemotaxis: Methods, Physiology and Clinical Implications." New York: Raven Press, p 1–24.

Williams LT, Synderman R, Lefkowitz RJ (1977): Specific receptor sites for chemotactic peptides on human polymorphonuclear leukocytes. Proc Natl Acad Sci USA 74:1204–1208.

Zigmond SH (1977): The ability of polymorphonuclear leukocytes to orient in gradients of chemotactic factors. J Cell Biol 75:606-616.

Zigmond SH (1978a): A new visual assay of leukocyte chemotaxis. In Gallin JI, Gallin PQ (eds): "Leukocyte Chemotaxis: Methods, Physiology and Clinical Implications." New York: Raven Press, pp 57–66.

Zigmond SH (1978b): Chemotaxis by polymorphonuclear leukocytes. J Cell Biol 77:269–287.

Zigmond SH, Levitsky HI, Kreel BJ (1981): Cell polarity: An examination of its behavioral expression and its consequences for polymorphonuclear leukocyte chemotaxis. J Cell Biol 89:585–592.

White Cell Mechanics: Basic Science and
Clinical Aspects, pages 195–208

Interactions Between Neutrophils and Microvascular Endothelial Cells Leading to Cell Emigration and Plasma Protein Leakage

T. J. Williams, P. J. Jose, M. J. Forrest, C. V. Wedmore, and G. F. Clough

Department of Pharmacology, Institute of Basic Medical Sciences, Royal College of Surgeons of England, Lincoln's Inn Fields, London WC2A 3PN, England

INTRODUCTION

A characteristic feature of the acute inflammatory response is the local accumulation of neutrophil leukocytes. Neutrophil accumulation in response to local injury was first described about 150 years ago as "lymph globules" pavementing the walls of microvessels. The neutrophils were then observed to penetrate the vessel walls and accumulate in the tissue affected by the inflammatory agent. These phenomena were first observed by Dutrochet [1824], Addison [1843], and Waller [1846] and studied in more detail by Cohnheim [1889]. The function of this process is to recruit blood cells capable of phagocytosis and microbial killing to a site of infection or injury; but how does the inflammatory stimulus cause neutrophil accumulation?

The mechanisms underlying neutrophil accumulation in a tissue are complex and are fundamentally dependent on the interaction between the neutrophil and the vascular endothelial cell, an interaction triggered by a chemical signal generated in the tissue. The mechanisms involved can be best understood by considering the acute sequence of events which occur following infection of a tissue with microorganisms. This sequence is outlined below.

RECOGNITION

The microbe has to be recognised as foreign before leukocytes can be recruited to eliminate it. The complement system in tissue fluid has an important role here. If antibodies to the microbe or its products are present, immune complexes can form and induce "classical pathway" complement activation. In addition, certain microbial cell walls and microbial products (eg, endotoxin) can induce complement activation in the absence of antibody, ie, "alternative pathway" activation. Complement activation results in coating (opsonization) of the microbes with C3b. This facilitates subsequent phagocytosis by neutrophils which have C3b receptors. The final product of activation is a macromolecular complex (C5b-9) thought to be in the form of a cylinder with a lipophilic end which can insert itself into the foreign cell causing lysis [Tranum-Jensen et al, 1978; Tschopp et al, 1982].

CHEMICAL SIGNALS TO THE MICROVASCULAR BED

Several protein cleavage by-products are produced as a consequence of complement activation. Probably the most important as a signal is C5a. Enzymic cleavage of the fifth component of complement is necessary to initiate the formation of the macromolecular lytic complex described above. The larger split product C5b combines spontaneously with C6, C7, C8, and C9. The smaller by-product is C5a, a polypeptide having a known sequence of 74 amino acids [Fernandez and Hugli, 1978]. C5a exhibits potent chemotactic activity (directed movement along a concentration gradient) and chemokinetic activity (increased random movement) for neutrophils in test systems in vitro.

Neutrophils are also attracted towards secretory products of microbes. Synthetic peptides resembling these products, eg, N-formyl-methionyl-leucyl-phenylalanine (FMLP), are very potent chemoattractants in vitro.

More recently it has been discovered that one of the family of leukotrienes, leukotriene B_4 (LTB_4), a 5-lipoxygenase product of arachidonic acid [Samuelsson and Hammarstrom, 1980], has potent chemotactic and chemokinetic activity for neutrophils [Ford-Hutchinson et al, 1980]. LTB_4 is released in certain anaphylactic reactions and is also released from neutrophils stimulated with calcium ionophore.

Thus, endogenous chemical signals can be generated in extravascular tissue fluid in response to microbes (eg, C5a), exogenous substances secreted by microbes can act in the same way (eg, formyl peptides) and neutrophils can themselves secrete signalling substances (eg, LTB_4) which may be important for sustaining neutrophil recruitment.

ADHERENCE OF NEUTROPHILS TO ENDOTHELIAL CELLS

A key phase in neutrophil recruitment, upon which all subsequent events depend, is neutrophil adherence. Some indications of the mechanisms involved can be obtained by intravital microscopy, eg, of exposed mesentery in animal experiments. In capillaries, blood cells travel in single file, a common observation being columns of erythrocytes preceded by single leukocytes. On entering the wider diameter postcapillary venules, the leukocytes tend to be pushed toward the vessel wall by the columns of smaller erythrocytes. In response to an inflammatory stimulus or when a chemoattractant is applied, increased adherence between neutrophils and venular endothelial cells is apparent. This results in neutrophils rolling along the wall and becoming fixed with their surfaces adjacent to the wall becoming flattened. These phenomena were studied in a quantitative intravital microscopic study of rat mesentery and hamster cheek pouch by Atherton and Born [1972]. To what extent these events occur normally in the absence of an inflammatory stimulus is difficult to establish, since exposure of a microvascular bed for intravital microscopy causes some tissue injury.

Precisely how substances like C5a can induce adherence is not known. Incubation of neutrophils with these substances in vitro causes an increased adhesiveness as evidenced by neutrophil aggregation. Thus, it is possible that during their passage along capillaries and venules, neutrophils experience a surface change because of diffusion of chemoattractant into the microvessel lumen. Collision between neutrophils and venular endothelial cells then results in adherence, low adhesion and high shear force giving rolling, and higher adhesion giving immobilisation of neutrophils on the vessel walls. That sticking occurs preferentially in small venules in most tissues (although in the lung, capillaries are affected), implies a specialisation of endothelial cells in these vessels; ie, that endothelial cell surfaces are more receptive to sticky neutrophils. Alternatively, the chemoattractant may diffuse into the microvessel lumen and present itself on the endothelial cell surface. Yet another possibility is that the chemoattractant stimulates the ablumenal surface of the endothelial cell, which induces an increased adhesiveness of its lumenal surface.

Several experiments have been carried out using neutrophils and cultured vascular endothelial cell monolayers in vitro [Lackie and deBono, 1977], showing that the addition of chemoattractant increases adherence [Hoover et al, 1978; Smith et al, 1979]. Increased neutrophil adhesion to artificial surfaces, eg, nylon fibres, has also been demonstrated in the presence of chemoattractant [MacGregor et al, 1974]. These experiments, however, do not resolve the question of the relative active participation of neutrophils and endothelial cells in adherence. Recent experiments in which endothelial cells were pretreated with chemoattractants and washed before addition of neutrophils showed no increased adhesion [Tonneson et al, 1982]. On the other hand, addition of the substances in the presence of both neutrophils and endothelial cells resulted in enhanced adhesion. This suggests that the neutrophil is the active participant. The situation in vivo may be different, however, as it has been observed that neutrophils dislodged from endothelial cells at a site of injury do not stick further downstream, which has been interpreted to show active participation of specialised venular endothelial cells.

PASSAGE OF NEUTROPHILS THROUGH ENDOTHELIAL CELL JUNCTIONS

The passage of neutrophils through the vessel wall was studied in detail in an electronmicroscopic study by Marchesi and Florey [1960] and Florey and Grant [1961]. A clear pseudopodium was seen emerging from the flattened surface of the neutrophils adjacent to the endothelial cell. Once through the endothelial cell layer, the pseudopodium was followed by the rest of the neutrophil. Frequently, neutrophils were observed sandwiched between endothelial cells and basement membrane. The endothelial cells appeared thin and under tension at this stage but junctions, which were often above the neutrophils, appeared to be closed. The site of penetration was described as at, or near, a junction between endothelial cells, although few observers now consider neutrophils to pass through individual endothelial cells.

How a chemoattractant induces neutrophil extravasation is not known. That neutrophils, once adhered, pass through junctions in endothelial cell monolayers in vitro in the absence of chemoattractants [Gordon and Pearson, 1980] suggests that the chemoattractant is not necessary for this stage. In vivo, however, a chemoattractant gradient at junctions may be necessary. A reverse gradient, with chemoattractant present intravascularly, is reported to cause neutrophil/endothelial cell adherence with no emigration [Henson et al, 1982], at least in the lung. It is still conjectural whether a cement substance is present at endothelial junctions and whether this substance is digested by neutrophils during their passage [Marchesi and Florey, 1960].

PASSAGE OF NEUTROPHILS THROUGH BASEMENT MEMBRANE

Marchesi and Florey [1960] described neutrophils being held up under the basement membrane and pericytes during neutrophil extravasation. It is possible that chemoattractants fix to basement membrane and that the neutrophil is stimulated to secrete proteolytic enzymes at the point of contact between the leukocyte and basement membrane. The neutrophil could then move through the hole digested by the enzymes.

Once outside the vessel wall, the neutrophil can move under the influence of the concentration gradient towards the site of generation of the chemoattractant. Subsequent contact with the infecting agent can then be followed by phagocytosis by the neutrophil.

THE LINK BETWEEN NEUTROPHIL EMIGRATION AND INCREASED MICROVASCULAR PERMEABILITY

Cohnheim observed a parallel between leukocyte accumulation and leakage of plasma leading to tissue oedema in frog microvessels responding to local injury [Cohnheim, 1889]. Thoma suggested that leukocyte emigration caused injury to the blood vessel wall resulting in an increase in its permeability; this then led to further leukocyte emigration [Thoma, 1896]. Studies on inflammatory responses in animals, eg, the Arthus reaction, showing a suppression of oedema formation by depletion of circulating neutrophils, further reinforced the idea of a connection between neutrophils and increased vascular permeability [Humphrey, 1955]. It was suggested that phagocytosing neutrophils leak enzymes which damage the blood vessel walls [Cochrane, 1967]. This was associated with the later phases of the acute reaction when neutrophils are apparent in high numbers in the tissues. In some experiments, however, it was reported that phases of neutrophil accumulation could be dissociated from phases of increased permeability, and the suggestion was made that the two phenomena were not related [Hurley, 1963].

Our interest in this problem developed from a pharmacological study of endogenous chemical mediators responsible for inducing local oedema in an inflammatory model in rabbit skin [Williams and Peck, 1977]. Intradermal injection of killed *Bordetella pertussis* organisms, boiled yeast cells (zymosan), or insoluble immune complexes was found to result in acute local oedema formation with a peak rate of protein leakage (as measured using intravenous ^{125}I- or ^{131}I-albumin) at 1–2 hours and a duration of 3–5 hours. It was discovered that oedema formation in this model results from two mediators acting synergistically. One of these mediators increases venular permeability to plasma proteins, but this produces little oedema in the absence of the second mediator, a prostaglandin (PG), which dilates arterioles and increases local blood flow. The exact mechanism whereby prostaglandins potentiate oedema remains undetermined, but elevation of hydrostatic pressure within the venule lumen is probably important. In support of this, another potent vasodilator, vasoactive intestinal polypeptide, is also a potent oedema potentiator [Williams, 1982].

We became interested in the identity of the substance responsible for increased venular permeability. Histamine and bradykinin appeared not to be important because antihistamines and kinin-formation inhibitors had little effect on oedema formation. We then found that incubation of blood plasma with the inflammatory stimulus (zymosan, etc) resulted in the generation of a permeability-increasing factor [Williams, 1978]. These plasma samples only induced oedema if mixed with a vasodilator prostaglandin before intradermal injection. Subsequent purification and characterisation showed the plasma factor to be the complement-derived polypeptide, C5a [Williams and Jose, 1981]. Human C5a was also shown to have the same activity in rabbit skin [Jose et al, 1981].

The long-established link between complement activation and increased microvascular permeability is via release of mast cell histamine by the "anaphylatoxins," C5a and C3a. The responses we observed, however, were not abolished by antihistamines [Williams and Jose, 1981]. Further, removal of the carboxyl-terminal arginine from human C5a by carboxypeptidase B (to produce C5a des Arg) abolished histamine-releasing activity, but not permeability-increasing activity [Jose et al, 1981].

These observations led us to investigate the mechanism of action of C5a in rabbit skin [Wedmore and Williams, 1981a]. Differences in time courses of responses in skin indicated that C5a increases microvascular permeability by a different mechanism from that of histamine and bradykinin. Responses to bradykinin and histamine were fast in onset, significant responses being detectable at 1.5 minutes, whereas there was a latent period of 6 minutes before responses to C5a were apparent. In addition, responses to bradykinin and histamine were of short duration ($t\frac{1}{2}$ = 4–6 minutes) whereas those to C5a were protracted ($t\frac{1}{2}$ = 90–100 minutes).

Initially we considered that acute responses, apparent within a few minutes, were unlikely to involve neutrophils; however, we were able to demonstrate that the permeability responses were entirely dependent on circulating neutrophils [Wedmore and Williams, 1981a]. Depletion of circulating neutrophils abolished responses to C5a in the skin but had no effect on responses to bradykinin and histamine (all agents mixed with PGE_2). Further, responses to C5a could be restored by cross blood transfusion between normal and depleted rabbits. In this study other chemoattractants, FMLP and LTB_4, were shown to exhibit activity in the skin similar to that of C5a. Both substances induced little oedema alone, but were potent when mixed with PGE_2 and again these responses were neutrophil dependent. In another study, where the accumulation of neutrophils was measured in skin, the time course of neutrophil accumulation was found to parallel that of plasma protein leakage in response to C5a [Issekutz, 1981]. Similar observations have now been made in another system, the hamster cheek pouch preparation, using LTB_4 as the chemoattractant [Bjork et al, 1982] although the participation of a vasodilator was not obligatory in this case presumably because of a higher basal blood flow.

These observations indicate that the release of chemoattractant extravascularly in a tissue can cause neutrophil accumulation which, in turn, induces a rapid elevation in microvascular permeability. The discrepancies between neutrophil accumulation and plasma protein leakage seen in some inflammatory reactions, as referred to earlier, can be explained by the release of two types of mediator capable of causing increased permeability; ie, direct-action mediators (eg, histamine and bradykinin) and neutrophil-dependent mediators (eg, C5a, LTB_4, and FMLP). It is also possible that mediators exist which cause neutrophil accumulation without increased microvascular permeability, but this awaits verification.

Histamine and bradykinin are thought to induce increased permeability by stimulating contraction of endothelial cells or pericytes [Wayland et al, 1975] and thus opening junctions; however, the mechanism whereby neutrophils increase permeability is not clear. There are several possibilities. The initial adherence of the neutrophil could induce endothelial cell contraction. The passage of neutrophils between endothelial cells could leave junctions open (although this was not observed in the electronmicrographs of Marchesi and Florey, 1960). Alternatively, the chemoattractant could induce the neutrophil to secrete a second mediator which acts on endothelial cells (perhaps causing contraction). This putative second mediator could be an oxygen radical, an enzyme, or a substance such as the phospholipid platelet-activating factor (PAF) which is thought to act directly on endothelial cells [Wedmore and Williams, 1981b].

Figure 1 shows an example of the synergism between microvascular permeability-increasing mediators, C5a and LTB_4, and the vasodilator mediator, PGE_2. Figure 2 shows an interpretation of the mechanisms underlying our model inflammatory reactions.

INCREASED MICROVASCULAR PERMEABILITY INDUCED BY THE PRESENCE OF INTRAVASCULAR CHEMOATTRACTANT

The above section is concerned with events occurring after extravascular complement activation. Under some circumstances intravascular

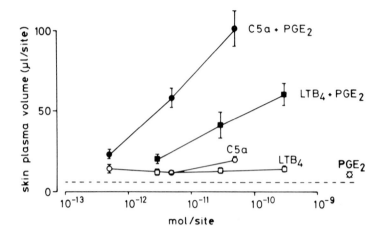

Fig. 1. Local oedema formation in the rabbit induced by intradermal injections of mixtures of C5a + PGE$_2$ and LTB$_4$ + PGE$_2$. Oedema formation was measured as the local accumulation of intravenously injected ^{125}I-albumin over a 30-minute period and expressed as μl plasma exudation per skin injection site. Responses are shown for three doses (mole/100-μl injection) of C5a tested alone (O) and with a fixed dose of PGE$_2$ (●). Similarly, responses are shown for three doses of leukotriene B$_4$ tested alone (□) and with PGE$_2$ (■). The fixed dose of PGE$_2$ used (3×10^{-10} mole/100 μl) induces little oedema formation when tested alone (◊). Note that C5a and LTB$_4$ induce little oedema formation when tested alone but both induce significant, dose-related oedema formation when tested in the presence of the vasodilator PGE$_2$. These oedema responses are entirely dependent on circulating neutrophils (unlike responses to histamine and bradykinin). The dashed line represents the skin plasma volume of sites injected with phosphate-buffered saline. Each point represents the mean ± SEM of six injected sites.

complement activation occurs. Neutropenia can be observed during haemodialysis, which is associated with pulmonary dysfunction. It has been suggested that dialysis membrane can activate intravascular complement by the alternative pathway leading to C5a generation, and that C5a can induce aggregates of circulating neutrophils which sequester mainly in the lung [Craddock et al, 1977]. This hypothesis has been extended to other situations where intravascular activated complement is thought to induce lung injury, eg, in adult respiratory distress syndrome (for review see Rinaldo and Rogers [1982]). Several lines of experimental evidence support the idea. Infusions of complement-activated plasma into the pulmonary artery was reported to induce increased microvascular permeability in the lungs of sheep and rabbits [Craddock et al, 1977]. Similarly, intravenous injection of a complement activator, cobra venom factor,

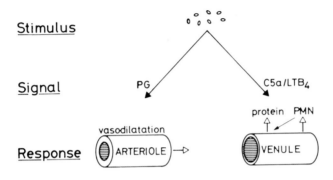

Fig. 2. A representation of an acute inflammatory response based on observations of the microvascular changes induced by a microbial stimulus (yeast cells) in rabbit skin. The stimulus induces the production of two signalling substances, a vasodilator prostaglandin (PGE_2 or PGI_2) and the complement-derived polypeptide C5a. The prostaglandin dilates arterioles increasing blood flow to the tissue. This has the secondary effect of causing increased hydrostatic pressure within the venule lumen resulting in passive venular distension. C5a regulates the adherence of neutrophils to venular endothelial cells followed by emigration. By an unknown mechanism this interaction between neutrophils and endothelial cell induces increased venular permeability to plasma proteins. In this model little plasma protein leakage occurs in the absence of the prostaglandin. This can be largely attributed to increased hydrostatic pressure within the venule lumen, although other additional mechanisms may contribute to the synergism between C5a and prostaglandins. Direct evidence for the generation of the two mediators has been obtained in this model. In other situations other chemoattractants, such as leukotriene B_4, may act in the same way as C5a.

These mechanisms probably underly a functional system in which the microvascular bed is adapted in order to supply blood proteins (eg, antibodies and complement components) and phagocytes to an infected tissue.

induced the same effect in rats [Till et al, 1982]. Again in the rat, infusions of complement-activated plasma into the aorta was observed, using intravital microscopy, to result in margination of neutrophils and collection of neutrophil aggregates in the mesentery [Hammerschmidt et al, 1981]. In this study, leakage of fluorescein-labelled albumin was observed from microvessels.

All these studies suggest that C5a is generated in plasma, neutrophils then become adhesive and either adhere to endothelial cells or form aggregates and lodge in microvessels. However, it has been reported that intravascularly injected radiolabelled C5a accumulates predominantly in the lung with, or without, the presence of circulating neutrophils [Webster et al, 1981]. Thus, it has been suggested that C5a can bind to endothelial cells in the lung to promote neutrophil adherence [Henson et al, 1982]. Henson and his colleagues [Henson et al, 1982] were unable to induce increased lung microvascular permeability in rabbits injected intravenously with cobra venom factor, zymosan-activated plasma or purified C5a, in contrast to the results of the other studies mentioned above. These workers suggest, therefore, that other factors are necessary in addition to C5a. They have reported that PGE_2 may be such an additional factor, which may relate to the observations of synergism between C5a and PGE_2 in the skin, as discussed in the previous section.

CONCLUSIONS

The neutrophil has the essential function of killing foreign organisms and removing these and other unwanted solid material from tissues. Neutrophils have to be supplied to an affected tissue in numbers compatible with the size of the stimulus (eg, number of microorganisms). This is achieved by sending chemical signals, such as C5a, to the microvascular bed. How these chemicals regulate neutrophil accumulation is the subject of intensive investigation and the interpretations of these studies are conflicting. One important question is whether vascular endothelial cells play a purely passive role in the initial adherence between neutrophils and endothelial cells. In the lung, adherence is between neutrophils and capillary endothelial cells, whereas in most other tissues adherence occurs preferentially in small (especially postcapillary) venules. This implies that the endothelial cells in these vessels where neutrophils adhere are specialised, but this does not necessarily mean an active role for the endothelial cell; specialised endothelial cells may provide a better surface for the attachment of a chemoattractant-stimulated neutrophil. Alternatively, blood flow characteristics in different regions of the microvascular bed may be important. One study showed the adherence of neutrophils to the walls of capillaries and arterioles as well as to the venules of rat mesentery when C5a was infused intravascularly [Hammerschmidt et al, 1981]. Interpreted together with other experimental findings, this is suggestive of an active role for venular endothelial cells when chemoattractant is generated extravascularly and that this selectivity of adherence does not occur when neutrophils are made more adherent by direct contact with chemoattractant. The results of Hammerschmidt et al [1981] could, however, be caused by neutrophil aggregates blocking microvessels and changing the blood-flow distribution within the tissue.

The distribution of intravenously injected ^{125}I-labelled C5a and C5a des Arg in rabbits, and the observation that neutrophil depletion did not affect distribution, have been interpreted as demonstrating specificity of binding to endothelial cells in different vascular beds. This is an interesting idea, and detailed studies of binding of C5a and C5a des Arg to endothelial cells should prove fruitful.

It is probable that neutrophil/endothelial cell adherence mechanisms triggered by chemoattractants have evolved for the purpose of responding to a local extravascular signal. If this is the case, the idea of active participation of endothelial cells is attractive. One possibility discussed by Tonnesen et al [1982] is diffusion of C5a from the tissue through endothelial cell junctions and fixation of the substance on the lumenal surface of the cell. Another idea, mentioned here previously, is stimulation of the ablumenal surface of the endothelial to induce a change in its lumenal surface. A third, speculative, idea could be added: C5a on the ablumenal surface of the endothelial cell could be translocated through the cell, perhaps bound to cell membrane.Vesicles or plasma membrane invaginations seen in endothelial cells by electronmicroscopy are objects much beloved by vascular physiologists as a basis for theories of vesicular transport of proteins out of blood vessels. Could it be that some of these intracellular objects represent devices for translocating molecules, such as C5a, in the opposite direction?

Inflammatory oedema has been generally regarded as a deleterious, secondary consequence of a local defence reaction in which protein leakage occurs because of injury to microvessels. Whilst protein leakage can occur as a consequence of microvascular injury, we have reasoned that inflammatory oedema in our model reflects the result of a functional system for supplying blood proteins to a tissue. The events occurring after a local microbial infection can be summarised as follows. Microorganisms infecting a tissue can activate complement in extravascular tissue fluid resulting in lysis (with some organisms) and opsonization for subsequent phagocytosis. As the amount of extravascular fluid is limited, this process would quickly deplete extravascular complement if a mechanism did not exist to supply more complement from the blood. In our model C5a is important for complement supply. C5a, a by-product of activation, controls protein leakage from microorganisms in proportion to the size of the stimulus. If antibody is present the same reasoning can be applied: Activation can be initiated more efficiently by antibody combined with the microorganism and protein leakage will supply both complement and further antibody from the blood.

Why, then, should neutrophils be involved in protein supply to the tissue? This can only be speculation, but the neutrophils are, in effect, facilitating the preparation (lysis and opsonisation) of the microbes for subsequent phagocytosis. Thus, in the early phase of a response, neutrophils in association with the vessel wall induce protein leakage and then later move out into the tissue to effect phagocytotis.

Whether this functional process should be regarded as causing damage to the vessel wall is a contentious point. Experiments have been performed showing damage of cultured endothelial cells in the presence of neutrophils and chemoattractants using ^{51}Cr release as an index of injury. However, the neutrophils may be responding in a manner analogous to their response to foreign cells because of changes in endothelial cells during culture.

It seems likely that chemoattractant is normally generated extravascularly. More rarely complement may be activated intravascularly, and under these circumstances protein leakage, especially from the lung, may be by an entirely different mechanism from that involved in extravascular generation. During intravascular activation (sometimes initiated by artificial interference with the circulation) the interaction between neutrophils and endothelial cells may well result in endothelial cell damage.

The microvascular endothelial cell is the "gate" cell controlling the influx of blood cells and proteins in response to an inflammatory stimulus. An understanding of this cell and its interaction with chemical mediators and leukocytes will be a key step in the understanding of the inflammatory process.

ACKNOWLEDGMENTS

We are grateful to the following for supporting the work of our laboratory: The Medical Research Council, Wellcome Trust, Arthritis and Rheumatism Council, Vandervell Foundation, Glaxo Group Research Ltd., Marks and Spencer, Ltd., and the George Clarke Bequest.

REFERENCES

Addison W (1843): Experimental and practical researches on the structure and function of blood corpuscles; on inflammation, and on the origin and nature of tubercules in the lungs. Trans Prov Med Surg Assoc 11:223–306.
Atherton A, Born GVR (1972): Quantitative investigations of the adhesiveness of circulating polymorphonuclear leucocytes to blood vessel walls. J Physiol 222:447–474.
Bjork J, Hedqvist P, Arfors K-E (1982): Increase in vascular permeability induced by leukotriene B_4 and the role of polymorphonuclear leukocytes. Inflammation 6:189–200.
Cochrane GC (1967): Mediators of the Arthus and related reactions. Prog Allergy 11:1–35.
Cohnheim J (1889): Lectures on General Pathology, Section 1. The pathology of the circulation. London: The New Sydenham Society.

Craddock PR, Fehr J, Brigham KL, Kronenberg RS, Jacob HS (1977): Complement and leukocyte-mediated pulmonary dysfunction in hemodialysis. N Engl J Med 296:769–773.

Dutrochet MH (1824): Recherches Anatomiques et Physiologiques sur la Structure Intime des Animaux et des Vegetaux, et sur leur Motilite. Paris: Bailliere et Fils.

Fernandez HN, Hugli TE (1978): Primary structural analysis of the polypeptide portion of human C5a anaphylatoxin. Polypeptide sequence determination and assignment of the oligosaccharide attachment site in C5a. J Biol Chem 153:6955–6964.

Florey HW, Grant LH (1961): Leucocyte migration from small blood vessels stimulated with ultraviolet light: An electron-microscope study. J Pathol Bacteriol 82:13–17.

Ford-Hutchinson AW, Bray MA, Doig MV, Shipley ME, Smith MJH (1980): Leukotriene B: A potent chemokinetic and aggregating substance released from polymorphonuclear leucocytes. Nature 286:264–265.

Gordon JL, Pearson JD (1980): The interaction of vascular cells and blood cells. Agents Actions [Suppl] 7:85–89.

Hammerschmidt DE, Harris PD, Wayland JR, Craddock PR (1981): Complement-induced granylocyte aggregation in vivo. Am J Pathol 102:146–150.

Henson PM, Larsen GL, Webster RO, Mitchell BC, Goins AJ, Henson JE (1982): Pumonary microvascular alterations and injury induced by complement activation, neutrophil sequestration and prostaglandins. Ann NY Acad Sci 384:287–300.

Hoover RL, Briggs RT, Karnovsky MJ (1978): The adhesive interaction between polymorphonuclear leukocytes and endothelial cells in vitro. Cell 14:423–428.

Humphrey JH (1955): The mechanism of Arthus reactions. 1. The role of polymorphonuclear leucocytes and other factors in reversed passive Arthus reactions in rabbits. Br J Exp Pathol 36:268–282.

Hurley JV (1963): An electron microscopic study of leucocyte emigration and vascular permeability in rat skin. Aust J Exp Biol Med Sci 41:171–186.

Issekutz AC (1981): Vascular responses during acute neutrophilic inflammation. Their relationship to in vivo neutrophil emigration. Lab Invest 45:435–441.

Jose PJ, Forrest MJ, Williams TJ (1981): Human C5a des Arg increases vascular permeability. J Immunol 127:2376–2380.

Lackie JM, deBono D (1977): Interactions of neutrophil granulocytes (PMNs) and endothelium in vitro. Microvasc Res 13:107–112.

MacGregor RR, Spagnuolo PJ, Lentneck AL (1974): Inhibition of granulocyte adherence by ethanol, prednisone, and aspirin, measured with a new assay system. N Engl J Med 291:642–646.

Marchesi VT, Florey HW (1960): Electron micrographic observations on the emigration of leucocytes. Quart J Exp Physiol 45:343–348.

Rinaldo JE, Rogers RM (1982): Adult respiratory-distress syndrome. Changing concepts of lung injury and repair. N Engl J Med 306:900–909.

Samuelsson B, Hammarstrom S (1980): Nomenclature for leukotrienes. Prostaglandins 19:645–648.

Smith RPC, Lackie JM, Wilkinson PC (1979): The effects of chemotactic factors on the adhesiveness of rabbit neutrophil granulocytes. Exp Cell Res 122:169–177.

Thoma R (1896): Textbook of General Pathology, and Pathological Anatomy. London: Adam & Charles Black.

Till GO, Johnson KJ, Kunkel R, Ward PA (1982): Intravascular activation of complement and acute lung injury: Dependence on neutrophils and toxic oxygen metabolites. J Clin Invest 69:1126–1135.

Tonneson MG, Smedly L, Goins A, Henson PM (1982): Interaction between neutrophils and vascular endothelial cells. Agents Actions [Suppl] 11:25–38.

Tranum-Jensen J, Bhakdi S, Bhakdi-Lehnen B, Bjerrum OJ, Speth V (1978): Complement lysis: The ultrastructure and orientation of the C5b-9 complex on target sheep erythrocyte membranes. Scand J Immunol 7:45–56.

Tschopp J, Muller-Eberhard HJ, Podack EF (1982): Formation of transmembrane tubules by spontaneous polymerization of the hydrophilic complement protein C9. Nature 298:534–538.

Waller A (1846): Microscopic examination of some principal tissues of the animal frame as observed in the tongue of the living frog, toad, etc. Lond Edin Dubl Phil Mag 29:271–287.

Wayland JH, Fox JR, Elmore MD (1975): Quantitative fluorescent tracer studies in vivo. Bibl Anat 13:61–64.

Webster RO, Larsen GL, Henson PM (1981): Tissue distribution of human C5a and C5a des Arg complement fragments in normal and neutropenic rabbits. Am Rev Respir Dis 123:41.

Wedmore CV, Williams TJ (1981a): Control of vascular permeability by polymorphonuclear leukocytes in inflammation. Nature 289:646–650.

Wedmore CV, Williams TJ (1981b): Platelet-activating factor (PAF), a secretory product of polymorphonuclear leucocytes, increases vascular permeability in rabbit skin. Br J Pharmacol 74:916–917.

Williams TJ (1978): A proposed mediator of increased vascular permeability in acute inflammation in the rabbit. J Physiol (Lond) 281:44–45.

Williams TJ (1982): Vasoactive intestinal polypeptide is more potent than prostaglandin E$_2$ as a vasodilator and oedema potentiator in rabbit skin. Br J Pharmacol 77:505–509.

Williams TJ, Jose PJ (1981): Mediation of increased vascular permeability after complement activation: Histamine-independent action of rabbit C5a. J Exp Med 153:136–153.

Williams TJ, Peck MJ (1977): Role of prostaglandin-mediated vasodilatation in inflammation. Nature 270:530–532.

White Cell Mechanics: Basic Science and
Clinical Aspects, pages 209–219
© 1984 Alan R. Liss, Inc., 150 Fifth Avenue, New York, NY 10011

Mechanical and Biochemical Aspects of Leukocyte Interactions With Model Vessel Walls

Larry V. McIntire and Suzanne G. Eskin

Biomedical Engineering Laboratory, Rice University, Houston, Texas 77251 (L. V. M.) and Department of Surgery, Baylor College of Medicine, Houston, Texas 77030 (S. G. E.)

INTRODUCTION

Circulating leukocytes normally must exit from the flowing blood in order to perform their bactericidal and immunological functions. The control system which allows leukocytes to recognize where to leave the vascular system and the details of adherence to and the passage through the endothelial lining are not completely understood. This chapter will concern itself with a method for examining leukocyte-vessel wall interactions in detail. Most of what is said about biochemical modulation will be applicable only to the subclass of polymorphonuclear neutrophil leukocytes (PMNL), though the system described is applicable to visualization of lymphocyte, monocyte, basophil, and eosinophil adherence to endothelial cells as well. Several very good reviews of PMNL-endothelial cell adhesion under static conditions are available [Hoover et al, 1978; Pearson et al, 1979; Lackie and Smith 1980; Hoover and Karnovsky, 1982; Lackie, 1982]. It is, however, well recognized that forces exerted by flowing blood on cells near the vessel wall play a crucial role in the success or failure of leukocyte-endothelium adhesive events. Low flow regions, such as postcapillary venules, are the primary site of PMNL margination in vivo. The method given below allows direct observation and quantitation of cell-vessel wall interactions in flowing whole blood or leukocyte suspensions, with the local fluid mechanical forces known and controlled. In addition, each cellular population can be individually manipulated biochemically to investigate the molecular mechanisms of endothelial cell-leukocyte adherence.

VISUALIZATION OF ADHERENCE UNDER FLOW

The ideal system in which to do studies of PMNL margination is to use the human animal in vivo. Normally this is ruled out by human experimentation committees; one can either choose another animal model or use human material in vitro. Both methods have severe drawbacks: species specificity and lack of control of many physical and biochemical variables versus a nonphysiological preparation. We believe significant knowledge can be gained using in vitro methods. An in vitro system should allow the use of whole blood with control of the local fluid mechanics. The forces acting on cells adherent to a vessel wall are (1) hydrodynamic drag, (2) normal forces (pressure), and (3) adhesive forces between the cell and substrate. The local fluid mechanics bring about number one above, mainly through the fluid shear stress acting over the exposed cell area. The adhesive force depends on the area of contact and the strength of interaction between cellular receptors. The magnitude of this force would be very sensitive to the biochemical state and morphology of the PMNL and the endothelial cells.

Labeling of PMNL

In order to visualize leukocyte-vessel wall interactions in whole blood, the PMNL were labeled and epifluorescent video microscopy employed. The antimalarial drug quinacrine dihydrochloride (Quin) was used as the fluorescent label. It is actively taken up into PMNL granules and does not label the nuclear material (Fig. 1). Labeling concentrations are $1-10~\mu M$. At these concentrations Quin does not affect PMNL chemiluminescence, chemotaxis, adherence to nylon fibers, or granule release [McIntire et al, 1982; Adams et al, 1983a,b]. The excitation wave length is 440 nm with an emission maximum of 505 nm, allowing the use of standard fluorescein isothiocyanate (FITC) filters if necessary. Labeling of erythrocytes is negligible at these concentrations, or at least any fluorescence from red cells is quenched by intracellular hemoglobin. Interestingly, Quin does label the nucleus of lymphocytes, allowing their visualization if desired. If excess Quin is used, endothelial cells will also become fluorescent.

Flow Chamber

Rheological factors have an important effect on cell-vessel wall interactions. Knowledge of the experimental wall shear stress is vital, if adherence studies in one laboratory are to be repeated in others and are to be applied to the in vivo situation. One convenient geometry where the fluid mechanics is well known, optical problems are minimal, and surfaces can be easily coated with proteins or cultured endothelial cells is the parallel plate flow chamber [Hochmuth et al, 1972; Richardson et al, 1977]. A

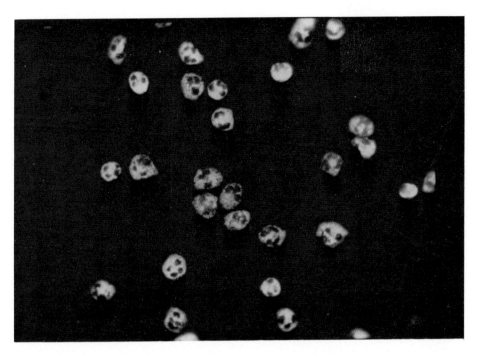

Fig. 1. Quinacrine-stained human polymorphonuclear leukocytes under fluorescent exitation. Note the localization of Quin in the granules and exclusion from the polymorphonuclear neutrophil leukocytes (PMNL) nucleus.

schematic of the device is given in Figure 2. The spacing between the plates is determined by the thickness of the silastic gasket but it is usually made to be 200–400 μm. The vacuum slot is used to hold the two flat surfaces in a fixed geometry without leakage under flow. If z is the flow direction and x the coordinate normal to the flat plate, with the origin at the center line, then at low Reynolds number the velocity profile is given by

$$V_z = \frac{3}{2} U_0 \left(1 - \left(\frac{x}{a} \right)^2 \right).$$ (1)

The wall shear stress, τ, can be calculated from

$$\tau = \mu \left. \frac{dV_z}{dx} \right|_{x=a} = \frac{3U_0\mu}{a} \simeq \frac{3\mu Q}{2a^2b}$$ (2)

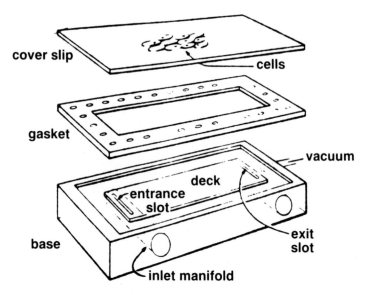

Fig. 2. Schematic of the parallel plate flow chamber, modified after Richardson et al [1977]. The gap spacing is determined by the gasket thickness. The vacuum slot is utilized to prevent fluid leakage under flow. Standard gaps are 200 μm to 400 μm.

where

$$a = \text{half height of the chamber,}$$

$$b = \text{channel width,}$$

$$U_0 = \text{average velocity,}$$

$$Q = \text{volumetric flow rate} = U_0 2ab,$$

$$\mu = \text{fluid viscosity,}$$

$$\left.\frac{dV_z}{dx}\right|_{x=a} = \text{wall shear rate.}$$

Thus given the volumetric flow rate, chamber dimensions, and the fluid velocity, one can easily determine the wall shear stress, and therefore the hydrodynamic forces exerted on attached cells. The restriction on the use of this equation is that $a \ll b$, since the velocity profile given above is not valid near the gasket (side) walls of the chamber. In practice this is usually true, since small values of a are used to minimize blood volume requirements. The wall shear rates attainable are 0–5,000 sec^{-1}, the range of physiological interest.

Model Vessel Walls

The flat plates can be easily coated with various proteins (such as collagen, fibrinogen, fibronectin) from solution or they can be used as a substrate for culture of human or animal endothelial cell monolayers. We normally use human umbilical vein or bovine thoracic aorta endothelial cells. For human cells, the culture procedure is a slightly modified form of that described by Gimbrone et al [1979]. Collagen-coated chambers can be used as a model for a damaged vessel wall.

Epifluorescent Video Microscopy Flow System

The flow chamber, with surfaces coated with either specific proteins or endothelial cells, is placed on a temperature-controlled epifluorescent microscope stage. It is then connected to the flow loop shown in Figure 3. This system is a constant pressure drop loop, with a variable resistance to introduce pulsatility, if desired. The volumetric flow rate is determined by the height of the upper reservoir above the stage, and is directly measured with an electromagnetic flowmeter. The upper reservoir has two outlets— one to the flow chamber and an overflow bypass directly to the lower reservoir. The flow from the return line (through the roller pump) is maintained slightly greater than the flow through the chamber, always assuring some overflow in the upper reservoir, and thus a constant pressure drop across the flow chamber. The volume of the loop has been made as small as 8 ml. Experiments can be run on time scales of minutes or hours. Events are recorded on videotape using a video camera with image intensifier, either in real time or using a time-lapse video recorder.

When analyzing videotapes of cell-cell interactions one can distinguish between three events: (1) collision, (2) contact, and (3) adhesion. The impingement of a leukocyte on the surface is considered a collision, and, for empirical reasons having to do with the time resolution of our video system, is defined as not to last longer than 30 msec. After collision, the contact phase of the adhesive event lasts up to half a second. In this interval, the leukocyte is stationary on the surface and either forms a "stable" bond and remains on the surface or is dislodged and is swept away by the flowing blood [Adams et al, 1983b]. The term adhesion, then, refers to cell-surface interactions which last for periods longer than 0.5 seconds. With PMNL, adherence is a dynamic quantity in this in vitro system, as it is in vivo [Schmid-Schoenbein et al, 1975, 1980]. Leukocytes will adhere, move locally for a period of time, and then detach or roll to move downstream. From the videotapes, one can determine number of adherent

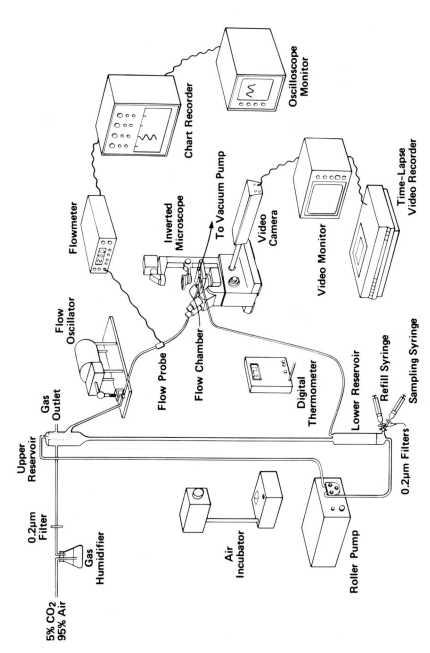

Fig. 3. Schematic of the overall flow loop. The pressure drop across the system is determined by the relative heights of the two reservoirs. The (constant) flow is monitored with a electromagnetic flowmeter. The entire assembly is maintained at 37°C using an air curtain.

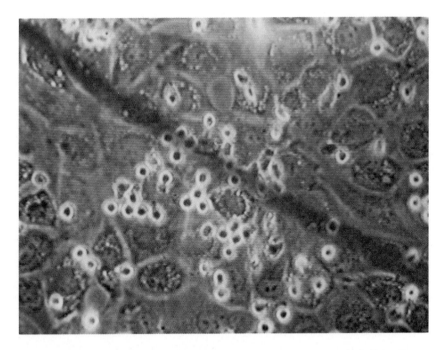

Fig. 4. Human PMNL being perfused over human umbilical vein endothelial cells at a wall shear stress of 1 dyne/cm². The smaller birefringent cells are the adherent PMNL. Nonadherent PMNL cannot be seen in this still photograph taken from the video monitor. The adherence process is very dynamic, with cells exchanging between the circulating and marginated pools.

PMNL as a function of time at fixed wall shear stress. A typical photograph from videotape is shown in Figure 4. From this morphological data, the kinetics of the adhesion process can be determined. The effects of fluid mechanical forces on PMNL-endothelial cell interaction can be investigated by repeating the experiment at different flow rates (and thus different wall shear stresses). In addition, the movement of PMNL through and under the endothelial cell monolayer can be observed.

BIOCHEMICAL MODULATION OF PMNL—SURFACE INTERACTION

The biochemical control mechanism of leukocyte adhesion to endothelial cells is still an active area of investigation. Arguments persist as to whether the endothelial cell plays an active or passive role in the process. Chemical modulators which may be important include (1) divalent cations, (2) arachidonic acid metabolites, (a) leukotrienes (B_4), (b) thromboxane A_2, (c) prostaglandins (PGI_2), (3) endotoxins and synthetic peptides N-formylmethionylleucine phenylalanine (FMLP), (4) complement fragments ($C5_a$), (5) superoxide anion, peroxides, hydroxyl radicals, (6) thrombin, and (7) platelet release products (serotonin).

The transient neutropenia seen in vivo during hemodialysis is due to margination of complement-activated PMNL, primarily in the pulmonary capillary beds [Toren et al, 1970; Jensen et al, 1973; O'Flaherty et al, 1978; Craddock, 1982]. A similar type of complement activation has been documented for patients undergoing cardiopulmonary bypass [Chenoweth et al, 1981].

The role of arachidonic acid metabolites in PMNL-endothelial cell interactions is complex. Endothelial cells seem to prefer to make prostacyclin via the cyclooxygenase-prostacyclin synthetase pathway, whereas PMNL appear to use the substrate to make leukotrienes via the 5 or 12 lipoxygenase pathway. To make matters more difficult, platelets normally utilize arachidonic acid to produce thromboxane A_2, via cyclooxygenase and thromboxane synthetase. PMNL, endothelial cells, and platelets all seem to have complete sets of enzymes to produce any of these three potent bioactive substances [Moncada and Vane, 1982], and inhibition of the normal preferred pathway may lead to shunting of arachidonic acid through a normally latent route. For example, stimulated cyclooxygenase-inhibited platelets (aspirin treated) may produce 12-hydroperoxy-5,8,10,14-eicosatetraenoic acid (12-HPETE) via lipoxygenase. This metabolite can then be utilized by PMNL to produce 5,12-dihydroxy eicosatetraenoic acid (5,12-DHETE), one isomer of which is the potent chemotractant leukotriene B_4 [Adams et al, 1982].

Superoxide anion, peroxides, and hydroxyl radicals can all be produced by activated PMNL and can be extremely cytotoxic to endothelial cells if the PMNL are adherent. Both the adherence of complement-stimulated granulocytes to endothelium and the subsequent leukocyte-induced endothelial cell damage can be enhanced by platelet release products [Boogaerts et al, 1982]. The mediator is believed to be serotonin.

Synthetic peptides (eg, FMLP) and bacterial endotoxins will also greatly increase PMNL-endothelial cell interaction [Hoover and Karnovsky, 1982]. Both the PMNL and endothelial cells have FMLP receptors. Adherence of untreated PMNL can be greatly increased by pretreatment of endothelial cells with this peptide—indicating an active role of the endothelial cells in this interaction. Similar phenomena may be very important in vivo in letting the circulating PMNL know where in the vasculature a bacterial invasion is underway, and therefore where they should marginate.

SUMMARY

Leukocytes in flowing blood are continually undergoing collisions with the blood vessel walls. Whether these collisions result in adherence depends on a delicate balance between the fluid mechanical drag force, which tends to dislodge the PMNL, and the adhesive force generated at the area of contact with the endothelium. Local blood flow rate controls the first of these forces, with the important parameter being the velocity gradient at the wall (wall shear rate). The detailed morphology of the endothelial cell and the PMNL upon collision and the biochemical state of these cells determine the adhesive force, because this force is a product of the strength of interaction times the area of contact. If a leukocyte flattens out or spreads on the vessel wall surface, it will reduce the hydrodynamic drag and increase the area of contact, leading to a more stable adhesion.

In the normal circulation, a significant fraction of the PMNL appear to be attached to the endothelium, particularly in the low flow venules. This leukocyte fraction is what hematology texts refer to as the marginal pool.

The adherent PMNL are a dynamic population in the sense that some return to the circulation, some move through the endothelial cell monolayer into the extravascular space, and others reman "attached" to the endothelial cells but roll along the surface in the flow direction. An equilibrium number is maintained under normal conditions by recruitment from the circulating pool.

Increased blood flow rate will increase the hydrodynamic force, tending to dislodge the PMNL, but will also (at least in large vessels) increase the number of collisions of circulating PMNL with the vessel wall by increasing the effective "diffusion" coefficient of the leukocyte. Thus if one studied the kinetics of PMNL adhesion on an initially clean endothelial monolayer surface under flow, increasing the whole blood flow rate would probably increase in initial rate of attachment but may decrease the equilibrium number of adherent leukocytes (Fig. 5).

The degree of cooperation in vivo between PMNL and endothelial cells required to enable the leukocytes to perform their bactericidal functions is fascinating. This type of interdependency also appears to be necessary in the inflammatory response [Wedmore and Williams, 1981]. Fluid mechanical forces in addition to biochemical events are crucial in regulating these interaction processes in vivo. In vitro models for examining PMNL adhesion should include control of the local fluid mechanics. The epifluorescent video microscopy system presented above gives one solution, which allows direct visualization of events, quantitation through image analysis of videotapes, and flexibility in biochemical manipulation of each of the cellular populations of interest independently.

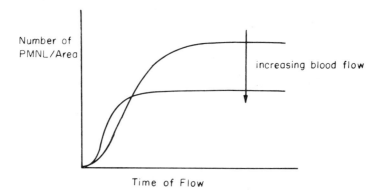

Fig. 5. Prediction of the effect of increasing whole blood flow rate on number of adherent PMNL per unit area of endothelial cell monolayer. When perfusion begins, there is an initial lag, then a rapid increase in cell adherence. The net rate of adherence gradually slows, with cell density reaching an equilibrium value. This "steady-state" density of adherent cells depends on the blood flow rate and the state of activation of the PMNL and the endothelium.

ACKNOWLEDGMENTS

This work was partly supported by grant HL-17437 from the National Heart Lung and Blood Institute of the National Institutes of Health, and Grant C938 from the Robert A. Welch Foundation.

REFERENCES

Adams GA, McIntire LV, Martin RR, Olsen JD, Sybers H (1982): The effects of heparin and polymophonuclear neutrophil leukocytes on platelet aggregate formation on collagen-coated tubes. Trans Am Soc Artif Intern Organs 28:444–448.

Adams GA, Putman M, McIntire LV, Martin RR (1983a): Quinacrine: An *in vivo* releaseable fluorescent label for polymorphonuclear neutrophil granules. J Lab Clin Med (in press).

Adams GA, Brown SJ, McIntire LV, Eskin SG, Martin RR (1983b): Kinetics of platelet and thrombus growth. Blood 62:69–74.

Boogaerts MA, Yamada O, Jacob HS, Moldow CF (1982): Enhancement of granulocyte endothelial cell adherence and granulocyte-induced cytotoxicity by platelet release products. Proc Natl Acad Sci USA 79:7019–7023.

Chenoweth DE, Cooper SW, Hugli TE, Stewart RW, Blackstone EH, Kirklin JW (1981): Complement activation during cardiopulmonary bypass. N Engl J Med 304:497–503.

Craddock PR (1982): Complement-Mediated Intravascular leukostasis and endothelial cell injury. In Nossel HL, Vogel HJ (eds.): "Pathobiology of the endothelial cell" New York: Academic Press pp. 369–389.

Gimbrone MA, Cotran RS, Folkman J (1979): Human vascular endothelial cells in culture. Proc Nat Acad Sci 76:5674–5679.

Hochmuth RM, Mohandas N, Spaeth EE, Williamson JR, Blackshear PL, Johnson DW (1972): Surface adhesion, deformation and detachment at low stress of red and white cells. Trans Am Soc Artif Intern Organs 18:325–332.

Hoover RL, Briggs KT, Karnovsky MJ (1978): The adhesive interaction between polymorphonuclear leukocytes and endothelial cells *in vitro*. Cell 14:423–428.

Hoover RL, Karnovsky MJ (1982): Leukocyte-endothelial interactions. In Nossel HL, Vogel HJ (eds): "Pathobiology of the Endothelial Cell." New York: Academic Press, pp 357–368.

Jensen DP, Brubaker LH, Nolph KD, Johnson CA, Nothum RJ (1973): Hemodialysis coil induced transient neutropenia and overshoot neutrophilia in normal man. Blood 41:399–408.

Lackie, JM (1982): Aspects of the behavior of neutrophil leukocytes. In Bellairs R, Curtis A, Dunn G (eds): "Cell Behavior." Cambridge: Cambridge University Press, pp 319–348.

Lackie JM, Smith RPC (1980): Interaction of leukocytes and endothelium. In Curtis ASG, Pitts JD (eds): "Cell Adhesion and Motility." Cambridge: Cambridge University Press, pp 235–272.

McIntire LV, Adams GA, Eskin SG, Martin RR (1982): Leukocyte and platelet interaction with protein and endothelial cell coated model vessels. Clin Hemorheol 2:273–281.

Moncada S, Vane JR (1982): The role of prostaglandins in platelet-vessel wall interactions. In Nossel HL, Vogel HJ (eds): "Pathobiology of the Endothelial Cell." New York: Academic Press, pp 253–258.

O'Flaherty JT, Craddock PR, Jacob HS (1978): Effect of intravascular complement activation on granulocyte adhesiveness and distribution. Blood 51:731–739.

Pearson JD, Carleton JS, Beesley JE, Hutchip A, Gordon JL (1979): Granulocyte adhesion to endothelium in culture. J Cell Sci 38:225–235.

Richardson PD, Mohammad SF, Mason RG (1977): Flow chamber studies of platelet adhesion at controlled spatially varied shear rates. Proc Eur Soc Artif Organs 1:175–188.

Schmid-Schoenbein GW, Fung YC, Zweifach BW (1975): Vascular endothelial-leukocyte interaction. Circ Res 36:173–184.

Schmid-Schoenbein GW, Usami S, Skalak R, Chien S (1980): The interaction of leukocytes and erythrocytes in capillary and post-capillary vessels. Microvasc Res 19:45–70.

Toren M, Goffinet JA, Kaplow LS (1970): Pulmonary bed sequestration of neutrophils during hemodialysis. Blood 36:337–340.

Wedmore CV, Williams TJ (1981): Control of vascular permeability by polymorphonuclear leukocytes in inflammation. Nature 289:646–650.

White Cell Mechanics: Basic Science and
Clinical Aspects, pages 221–236

Analysis of Cell Egress in Bone Marrow

Richard E. Waugh, Lewis L. Hsu, Patricia Clark, and Alfred Clark, Jr.

*Department of Radiation Biology and Biophysics, University of Rochester
School of Medicine and Dentistry, Rochester, New York 14642 (R. E. W.,
L. L. H.), Department of Mathematics, Rochester Institute of Technology,
Rochester, New York 14623 (P. C.), and Department of Mechanical
Engineering, University of Rochester, Rochester, New York 14642 (A. C.)*

INTRODUCTION

The bone marrow supplies approximately 2.5 million red blood cells and
1.2 million white blood cells to the circulation every second. To get from
the hemopoietic space where the cells grow and differentiate, to the blood
sinuses, which are contiguous with the circulation, the cells must pass
through small pores in the endothelial wall separating the two compartments. These pores range in size from slightly less than 1.0 μm to about 3.0
μm in diameter and from 0.25 μm to about 1.0 μm in thickness. Little is
known about the factors which regulate and control this efflux of cells
through these pores into the circulation. This analysis has been undertaken
to determine what role physical parameters such as membrane and
cellular deformability, pore size, and pressure gradients may play in the
regulation of the egress process. In the first section we consider the passage
of a single cell through a pore of fixed dimensions. The estimates of the
time required for a cell to pass through the boundary are then used to
calculate the number of pores needed to satisfy the body's need for new
cells as well as the fraction of pores which might be expected to contain
white cells or red cells at any given time. Finally, the expressions for cell
flux are used to illustrate how different physical parameters might be used
to regulate and control the rate at which cells leave the marrow.

RETICULOCYTE EGRESS

We begin with the analysis of a reticulocyte passing through a small
pore because the mechanical properties of erythrocytes are well known

and it is possible to perform an accurate analysis. (The influence of white cells on total cell efflux is considered in later sections.) The approach used in the analysis was chosen after examining electron micrographs of reticulocytes passing through pores from the hemopoietic space into the blood sinus (Fig. 1). Invariably, the geometry of the portion of the cell in the sinus was spherical. This suggested to us that the egress might be driven by a hydrostatic pressure gradient. We undertook the analysis to determine what pressure would be required to force the cell through a pore of such dimensions and how much time it would take to complete the passage. We also investigated the effects of changing pore radius and differences in cellular mechanical properties on the pressure and time required for egress.

We have used a simplified geometry to model the process so that reasonable estimates of the quantities of interest could be obtained without excessive calculation (Fig. 2). In the unstressed state the cell is assumed to be made up of two flat disks connected at the edge. As the cell passes through the pore, the center of the upper disk is deformed into a spherical section. Material is drawn from the lower disk around the edge of the disks. The deformation is assumed to be axisymmetric and to occur at

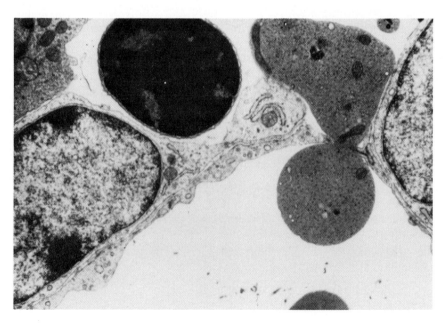

Fig. 1. Electron micrograph showing the spherical projection of a reticulocyte into the blood sinus during egress. [Taken from Lichtman et al, 1978, with permission.]

constant surface area [Evans et al, 1976]. The distance from the edge of the disk to the center of the pore decreases as more material is drawn through the pore. When the radius of the remaining disk equals the pore radius, egress is complete. It is assumed that the egress is driven by a constant pressure difference across the spherical contour in the blood sinus. The resistance to passage is due to the deformation of the planar membrane in the hemopoietic compartment. The interaction between the cell and the pore is assumed to be frictionless and the radius of the pore is constant. The volume of the cell is allowed to change freely, and the viscous resistance of the cytoplasm is neglected. (Several volume-conserving models have been attempted, but all have proven to be analytically intractable.)

This approach is similar to the one used by Evans [1973] to obtain the surface elastic shear modulus of red blood cell membrane from micropipette aspiration studies. Evans was able to calculate an unknown material property from a known aspiration pressure. We are able to estimate an unknown pressure from established material properties. The basic equation used in the analysis is the tangential force balance for an axisymmetric surface with no surface tractions [Evans and Skalak, 1979, Eq. 3.5.9]:

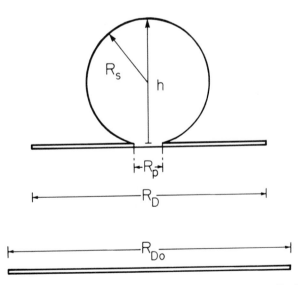

Fig. 2. The geometric model used in the analysis of reticulocyte egress. R_{Do} is the radius of the undeformed disk, R_D is the radius of the deformed disk, R_p is the radius of the pore, R_s is the radius of the spherical contour, and h is the height of the spherical projection.

$$r \frac{dT_m}{dr} = -2T_s, \tag{1}$$

where T_m is the tension (force/length) within the surface in the radial direction, T_s is the maximum shear resultant (force/length) in the surface, and r is the radial coordinate. T_s is related to the surface deformation via the constitutive equation for the surface [Evans and Hochmuth, 1976]:

$$T_s = \frac{\mu}{2} (\lambda^2 - \lambda^{-2}) + 2\eta V_s. \tag{2}$$

The total shear resultant in the surface is the sum of two terms. The first is the elastic contribution and the second is the viscous (dissipative) contribution. The material coefficients which characterize the membrane response to applied forces are the surface elastic shear modulus, μ, and the surface viscosity, η. The magnitude of the material deformation at any point is given by the material extension ratio, λ. The rate of material deformation is V_s.

Solution to the problem requires expressions for the material extension, λ, and the rate of material deformation, V_s, in terms of the radial coordinate in the deformed geometry, r. For an axisymmetric surface, $\lambda = r_0/r$, where r_0 is the radial coordinate of the material element in the undeformed geometry [Evans and Skalak, 1979, Eq. 5.3.7]. The relationship between r_0 and r is obtained by equating the deformed and the undeformed surface areas. Thus λ can be found for any position on the deformed surface, r, for given values of the area outside the pore, A_s, the pore radius, R_p, and the dimension of the undeformed disk, R_{Do}. The rate of deformation V_s for a planar surface is given by Evans and Skalak [1979, Eq. 2.5.26] as

$$V_s = \frac{dv_r}{dr}. \tag{3}$$

where v_r is the material velocity in the radial direction. Continuity requires that $v_r r =$ constant when the surface contour is flat. Therefore,

$$V_s = \frac{\dot{A}_s}{2\pi r^2}, \tag{4}$$

where $\dot{A}_s$ is the rate of area accumulation on the sinus side of the pore.

After substituting the expressions for λ and V_s as functions of r and geometric parameters into the combined form of Eqs. 1 and 2, we obtain a first-order differential equation for T_m as a function of r (see Eq. 6). Two boundary conditions are specified:

$$T_m = \frac{\Delta PR_s}{2} \qquad \text{at } r = R_p \qquad \qquad (5a)$$

$$T_m = 0 \qquad \text{at } r = R_D. \qquad \qquad (5b)$$

R_s is the radius of the spherical contour in the sinus and R_D is the instantaneous radius of the disks. The first condition presumes that the tension in the membrane is continuous from the sphere to the disk. The second condition is based on the fact that there is no deformation or rate of deformation on the lower disk ($T_s = 0$). Because we have constrained the geometry, it is impossible to satisfy the condition that $T_s = 0$ at the rim of the upper disk for nonstatic situations. This results in a discontinuity in T_s between the upper and lower disks. The choice of $T_m = 0$ is somewhat arbitrary; we could have chosen $T_\phi = 0$ at the rim. The latter choice results in longer times for egress. For a pore 1.0 μm in diameter the increase is about 5.0%, for a 1.5-μm pore the increase is about 15.0%, and for a 2.0-μm pore the increase is 25–30%.

The variables in the differential equation are separable:

$$dT_m = -\left[\mu(\lambda^2 - \lambda^{-2}) + \frac{2\eta \dot{A}_s}{\pi r^2} \right] \frac{dr}{r}. \qquad \qquad (6)$$

This equation is integrated over the upper disk from $r = R_p$ to $r = R_D$. Application of the two boundary conditions results in a functional relationship among the geometric parameters and the material constants. The solution is conveniently expressed in terms of the height of the sinus projection, h:

$$\frac{dh}{dt} = \left[\frac{\Delta P}{4h} (h^2 + R_p^2) - \mu \left\{ \ln \left(\frac{R_{Do}^2 - h^2}{R_{Do}^2} \right) + \frac{1}{2} \ln \left(\frac{h^2 + R_p^2}{R_p^2} \right) \right.\right.$$

$$\left.\left. + \frac{h^2}{2R_p^2} \right\} \right] \cdot \left[2\eta h \left(\frac{1}{R_p^2} - \frac{2}{2R_{Do}^2 - h^2} \right) \right]^{-1} \qquad A_s \leq A_{1/2} \quad (7a)$$

$$\frac{dh}{dt} = \left[\frac{\Delta P}{4h}\left(h^2 + R_p^2\right) - \mu\left\{\frac{2R_{Do}^2 - h^2}{2R_p^2} - 1 + \frac{1}{2}\ln\left(\frac{R_p^2}{2R_{Do}^2 - h^2 - R_p^2}\right)\right\}\right]$$

$$\cdot\left[2\eta h\left(\frac{1}{R_p^2} - \frac{2}{2R_{Do}^2 - h^2}\right)\right]^{-1} \quad A_s \geq A_{1/2} \quad (7b)$$

Two expressions are required because of the different relationships between λ and r for material originally in the upper or lower disk. $A_{1/2}$ is the area of one disk; when $A_s = A_{1/2}$ half of the area of the cell is in the blood sinus.

These expressions were used to calculate the pressure needed to form spherical projections of a given size as well as the time needed for egress. In the calculations we used material coefficients which have been measured on mature erythrocytes ($\mu = 0.006$ dyne/cm; $\eta = 0.0004$ dyne sec/cm). Preliminary measurements performed in our laboratory on circulating reticulocytes in the rabbit show little difference between the membrane material properties of reticulocytes and mature cells. The area of the cell was taken to be 140 μm^2. The pore radius and the pressure were varied as described.

For the static case (dh/dt = 0), Equations 7a and 7b can be used to obtain the equilibrium pressure needed to maintain a constant projection height (Fig. 3). As can be seen in the figure, this pressure reaches a maximum when the cell is approximately halfway out of the pore. (The maximum actually occurs slightly before the halfway point.) This peak represents the minimum pressure required for egress, ΔP_{min}. Figure 4 shows the strong dependence of ΔP_{min} on pore radius. For pores 1.0 μm in diameter and larger, the minimum pressure for egress is less than 2.2 × 10^3 dynes/cm^2 (less than 1.5 mm Hg). These values are small compared to the magnitudes of pressures measured in rabbit bone marrow by Michelsen [1967], who recorded differences in pressure on the order of 20.0 mm Hg between the marrow tissue and emissary veins just outside of the cortical bone. Based on these measurements it is reasonable to hypothesize pressure differences across the pore on the order of 2–5 mm Hg, sufficient to move reticulocytes through the boundary.

Equations 7a and 7b can be integrated numerically to obtain the time necessary for cell egress. Egress time is shown in Figure 5 as a function of driving pressure for different pore dimensions. The most interesting feature of the curves is the very short times needed to complete egress when ΔP_{min} is exceeded. For pores 1.0 μm in diameter and larger and pressures of about 3.0 mm Hg the time for egress is less than 0.3 second. This is an important result considering the fact that the pressures within the marrow are pulsatile [Michelsen, 1967] and pressures sufficient to

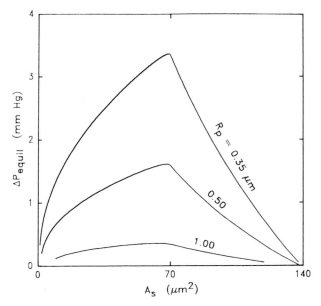

Fig. 3.   Equilibrium pressure, ΔP_{equil}, as a function of the area of the extruded sphere, A_s. The peak of the curve is the minimum pressure needed for egress.

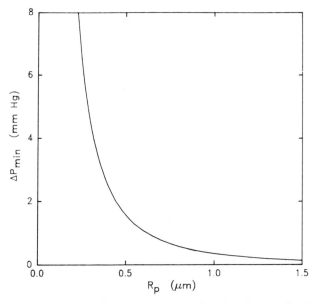

Fig. 4.   The minimum pressure for egress, ΔP_{min}, as a function of pore radius, R_p. Pressures become unrealistically large for pores less than 0.5 μm in diameter.

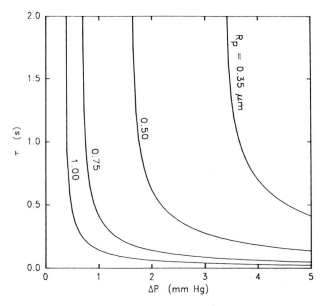

Fig. 5. The time needed to complete egress, τ, as a function of the pressure across the pore, ΔP, for different pore radii. For pressures only 12% greater than the minimum pressure for egress, the time for egress is less than 1.0 second.

cause egress may be transient. Dabrowski et al [1981] have shown a correlation between transient increases in pressure within the marrow and an increase in the release of reticulocytes into the circulation in rabbits, indicating that the pulsatile nature of the pressure within the marrow may contribute to cell release.

The effect of membrane properties on egress can be seen by inspection of Equations 7a and 7b. The minimum pressure for egress is directly proportional to the elastic shear modulus, and the time for egress is directly proportional to membrane viscosity.

Assessment of Spherical Contour

The following calculation was undertaken to test the validity of our assumption that the contour is perfectly spherical and to determine the minimum pressure needed to form a spherical contour. The approach was similar to the one used to calculate the surface contour when a tether is formed from a spherical vesicle [Waugh, 1982]. Two force balance equations and the elastic constitutive equation were used to obtain a differential equation, the solution of which gave the shape of the contour in the form of a functional relationship between a radial coordinate, r, and the direction of the surface normal given by an angle, θ. The material extension ratios were calculated assuming that the surface area was constant and the shape of the unstressed surface was planar. The calculated shape of the contour depended on the value of the pressure, ΔP, which was assumed to be constant and uniform over the surface. No significant deviation from the spherical geometry is observed for pressures as low as 10^{-3} mm Hg. This means that our geometric approximation is accurate, but because of the insensitivity of the shape to increases in pressure above 10^{-3} mm Hg, it is impossible to make further deductions about the magnitude of the pressure gradient across the pore from this calculation.

In addition to the shape of the contour, the analysis also yields values for the components of the surface force resultants. T_m is the resultant parallel to the z-axis and T_ϕ is the resultant in the circumferential direction. Two features of the calculated values of the resultants should be noted. First, the adequacy of the spherical approximation was confirmed by finding that the value for T_m calculated assuming a spherical contour is within 2.0% of the value for T_m calculated without assuming anything about the geometry of the surface contour. Second, values for T_ϕ are positive over most of the surface of the projection, but near the pore they can be negative (compressive). Because of the small bending stiffness of the membrane, these compressive resultants are likely to cause buckling and folding. Consistent with this result, scanning electron micrographs of cell projections in the sinus often show folds and wrinkles in the surface near the pore. Such folds would reduce the pressure needed for egress as well as the time of passage. In view of this fact, the values shown in Figures 3–5 should be regarded as upper-bound estimates.

Cytoplasmic Resistance

Throughout this treatment we have neglected the resistance of the cytoplasm to egress. Previous investigators who have neglected cytoplasmic resistance in dynamic red cell deformation [Evans and Hochmuth, 1976; Hochmuth et al, 1979] have justified their assumption by showing that the ratio of energy dissipation in the membrane to the dissipation in the cytoplasm is large. An equivalent, alternative approach is to show that the time for passage of a drop of cytoplasm in the absence of the membrane is much less than the passage of the membrane alone. The time for the passage of a fluid drop of volume V can be approximated by using the equation for creeping flow through a pore [Happel and Brenner, 1965, p. 153]

$$q = \frac{\Delta P \cdot R_p^{3}}{3\eta_c} \tag{8}$$

where q is the volumetric flow rate and η_c is the viscosity of the fluid. The time for a volume of fluid, V, to pass through the pore is

$$t_c = \frac{V}{q} = \frac{3\eta_c V}{\Delta P R_p^3}. \tag{9}$$

As an example, suppose the pore radius is 0.5 μm and the pressure difference is 2.0 mm Hg. The volume of the cell is 100 μm^3 and the cytoplasmic viscosity is 0.06 dyne sec/cm^2. In this case the passage time for the cytoplasm is about 0.05 seconds, compared to a membrane transit time of 0.6 seconds. Clearly, the membrane limits the rate at which the cell can pass through the pore.

WHITE CELL EGRESS

A detailed analysis of the passage of a white blood cell through a pore is impossible at this time because of the lack of knowledge about the mechanical properties of white blood cells undergoing large deformations. The fact that white blood cells are motile and the likelihood that motility plays a vital role in white cell egress raise significant questions about the physiological relevance of an analysis of passive white cell deformation. However, calculations of the flux of cells out of the hemopoietic space require some estimate of the time needed for a white cell to pass through the boundary. Because we know of no measurements of white cell transit times through pores we have made a crude estimate of the white cell transit time by modeling the cell as a viscous drop. A viscosity coefficient for the white cell cytoplasm has been measured by Schmid-Schönbein et al [1981] using micropipettes to produce small deformations of the cell. Using this value for viscosity ($\sim$100 dynes sec/cm^2), the equations for creeping flow through a pore, and assuming the cell has a volume of about 250 $\mu m,^3$ we calculate a passage time on the order of 100 seconds for a pore 2.0 μm in diameter at a pressure difference of 4.0 mm Hg (5.3×10^3 dynes/cm^2). This calculation is likely to be an overestimate of the transit time because Equation 8 assumes that the entire volume of fluid has the same viscosity, whereas in the case of a cell or a highly viscous drop the surrounding fluid has a considerably smaller viscosity. In spite of this, the time we calculate is similar to the times reported by Evans [this volume] and Cokelet [University of Rochester, personal communication] for aspiration of passive white cells into a micropipette. This time is considerably longer than the time it would take for a membrane to pass through a pore of similar dimensions (assuming the membrane properties were similar to the properties of a red cell membrane). In this case the cytoplasm limits the rate at which the cell can pass through the boundary. The nonspherical geometry of the white cell projections in the blood sinus indicates that the transpore pressure difference is not supported by a passive membrane. However, the nonspherical geometry does not allow us to distinguish between a cell which is moving actively and a passive cell with a highly viscous interior which is driven by a hydrostatic pressure.

The following discussion of cell fluxes at the endothelial boundary does not depend on the mechanism by which cells accomplish egress, but only on the time it takes to complete passage through the pore. It is possible (even probable) that cells in the active state could accomplish the passage in a much shorter time than passive cells, but because we know of no measurements of active passage of leukocytes through pores we can only speculate about how short the time might be. Because of this uncertainty a range of values for white cell egress time has been used in the calculations.

CELL FLUX AT THE BOUNDARY

Important insights about possible mechanisms for the regulation of cell egress can be obtained by applying the principle of mass balance to the flux of cells at the endothelial wall. For example, electron micrographic examination of bone marrow sections revealed that nearly all of the pores existing in the marrow are filled with cells [Lichtman et al, 1978]. This observation has been used as evidence for the hypothesis that pores only form when cells are in close contact with the hemopoietic face of the endothelial barrier. This conclusion presumes that if the pores were permanent structures a significant fraction would be empty at any given time. The principle of mass balance can be used to calculate the fractions of preexisting pores that would be filled or empty. (Cokelet [1981] and Skalak et al [1983] have used similar approaches in the analysis of the passage of cells through porous filters.)

Simply stated, the principle requires that under steady-state conditions the number of cells arriving at the boundary equals the number of cells leaving during any fixed period of time. Let N be the number of pores per unit area, f the fraction of pores that are filled, and τ the transit time of a cell through the boundary. The number of cells leaving the boundary per unit area per unit time is Nf/τ. The number of cells arriving at the boundary is given by the product of the total volume flux toward the boundary times the number of cells per unit volume in the hemopoietic space. The total volume flux is the product of volumetric flow rate through a single pore, q, times the number of empty pores per unit area, $N(1 - f)$. The number of cells per unit volume is equal to the volume fraction of cells,

H, divided by the volume of a single cell, V. The total flux balance can be written

$$\frac{Nf}{\tau} = \frac{HqN(1 - f)}{V}. \tag{10}$$

Equation 10 can be solved to obtain an expression for the fraction of filled pores,

$$f = \frac{1.0}{1.0 + \theta}, \tag{11}$$

where $\theta = V/Hq\tau$ is the ratio of the time a pore is empty to the time a pore is filled.

Neglecting the presence of white cells momentarily, we can calculate the fraction of filled pores for values characteristic of a population of red cells. The volume of the cell is about 100 μm^3. The hematocrit, H, is not known for the hemopoietic space, but a value of 0.5 seems reasonable. The flow through an empty pore, q, can be calculated from Equation 8, using the viscosity of plasma (0.016 dyne sec/cm^2). The time constant is obtained from the membrane analysis (Fig. 5). If the average pore diameter is 1.5 μm and the pressure difference across the pore is 4.0 mm Hg, the fraction of pores filled is 93%. This result does not disprove the hypothesis that pores only form when cells are in contact with the endothelium. However, it does show that the hypothesis is not supported by the fact that most pores in the marrow are filled, because nearly all the pores would be filled even if they were permanent structures.

If we account for the presence of white blood cells in the marrow, the fraction of filled pores is even higher. Let us suppose that there are two types of cells which can approach the boundary. The flux for each type is given by

$$Q_1 = \frac{Nf_1}{\tau_1} = \frac{NH_1q}{V_1}(1.0 - f_1 - f_2) \tag{12a}$$

$$Q_2 = \frac{Nf_2}{\tau_2} = \frac{NH_2q}{V_2}(1.0 - f_1 - f_2). \tag{12b}$$

The ratio of the flux of red blood cells to the flux of white blood cells is known for normal individuals in the absence of inflammatory stimuli. Using this data, and assuming that all cells compete on an equal basis for the same pores, we can deduce the ratio of the volume fractions of the two populations in the hemopoietic compartment. Taking the ratio of Equations 12a and 12b we find

$$\frac{Q_1}{Q_2} = \frac{H_1 V_2}{H_2 V_1}. \tag{13}$$

For an individual weighing 70 kg, the red cell flux, Q_1, is about 2.5×10^6 cells/sec, and white cell flux is about 1.2×10^6 cells/sec [Klebanoff and Clark, 1978, pp. 74–77]. The respective volumes are $V_1 \sim 100 \ \mu m^3$ and $V_2 \sim 250 \ \mu m^3$. Using Equation 13 we find the ratio of the cell volume fractions, $H_2/H_1 = 1.2$ If the volume fraction of reticulocytes in the marrow is 0.25, then the volume fraction of white cells would be 0.30. This result is in good agreement with the numbers of reticulocytes and granulocytes existing in the marrow measured by Donahue et al [1958].

This information can be used to calculate the fraction of pores containing cell type 1 or cell type 2. Solving Equations 12 for f_1 and f_2 we find

$$f_1 = \left[1.0 + \frac{\theta_1}{\theta_2} (1.0 + \theta_2) \right]^{-1} \tag{14a}$$

$$f_2 = \left[1.0 + \frac{\theta_2}{\theta_1} (1.0 + \theta_1) \right]^{-1}, \tag{14b}$$

where $\theta_1 = V_1/(\tau_1 H_1 q)$ and $\theta_2 = V_2/(\tau_2 H_2 q)$. Calculated values of f_1 and f_2 are shown in Table I for three different egress times for white blood cells. The values used for V_1, V_2, H_1, and H_2 were those described in the preceding paragraph. The time of red cell egress was taken to be 0.1 second. The value of q was calculated from Equation 8, using 0.75 μm for the pore radius, the viscosity of plasma (0.016 dyne sec/cm^2), and assuming a pressure difference of 3.0 mm Hg (4.0×10^3 dynes/cm^2). The

TABLE I. Calculated Values of f_1 and f_2 for Three Different Egress Times for White Blood Cells

τ_2 (sec)	f_1	f_2	$f_1 + f_2$	Pores/mm^3
1.0	0.169	0.812	0.981	1.5
10.0	0.020	0.977	0.998	12
100.0	0.002	0.998	>0.999	120

number of pores per mm^3 was calculated based on an estimated marrow volume of 1.0 liter. The results in Table I indicate that nearly all of the pores in the marrow should be filled with white cells if the egress time for white cells is 10.0 seconds or more. This prediction does not depend on the *mechanism* by which the cells leave the marrow (ie, actively or passively), but it does depend on the average time it takes a cell to pass through the pore. It also presumes the accuracy of the value calculated for the reticulocyte transit time, which could be in error if the pressure difference is near the minimum pressure for egress or if the pressure is pulsatile and sufficient for egress only a fraction of the time. The prediction, and the analysis used to obtain it, can be tested when the necessary experimental data become available. Examination of fixed bone marrow sections could provide an estimate of the actual number of pores in the marrow and the fraction which are filled with white cells or red cells. In vitro measurements could be performed to obtain an estimate of the egress time for white cells both in the presence and absence of chemotactic gradients.

Possible Physical Mechanisms for Controlling Flux at the Boundary

The expressions for the flux of cells at the endothelial boundary enable us to identify and evaluate potential mechanisms for the regulation and control of cell egress. The relative fluxes of white cells and red cells can be adjusted by changing the relative volume fraction of the cells (H_1 and H_2) in the hemopoietic space (Eq. 13). However, increasing H is not an effective means of increasing the *net* flux of cells out of the marrow. As can be seen from Equations 10 and 11, an increase in H requires an increase in the fraction of pores filled. Because nearly all of the pores are filled under normal conditions, increasing H can only increase the flux by a few percent. A simple and effective way to increase the flux is to increase the number of pores per unit area, N. Doubling N would double the flux without affecting the fraction of pores filled (assuming that the reservoir of cells in the hemopoietic space is not depleted).

Other possible mechanisms for increasing cell flux include increasing the driving pressure, ΔP, or the pore radius, R_p. These mechanisms work by decreasing the time for egress, τ, and increasing the flow per pore, q. (Note that these mechanisms may not work for actively migrating cells because the active process may be insensitive to pore size or pressure difference.) To evaluate the effect of pore size and pressure difference on the fraction of pores filled it is necessary to know how the product, τq, changes with these quantities. The flow rate, q, is proportional to $\Delta P R_p^3$ (Eq. 8). The dependence of τ on these quantities is more complex, but an order of magnitude analysis indicates that τ is inversely related to

ΔPR_p minus a constant. The constant represents the elastic threshold (ΔP_{min} or R_{pmin}), which must be overcome for egress to occur. The product goes as

$$\tau q \sim \frac{\Delta PR_p^{\,3}}{\Delta PR_p - \text{const.}}. \tag{15}$$

Increasing ΔP will decrease this product, causing a decrease in the fraction of filled pores. If the cells are near to the threshold for egress $R_p < 1.5\ R_{pmin}$) increasing R_p will have a similar effect, but if the cells are far from the threshold ($R_p > 1.5\ R_{pmin}$) increasing R_p would tend to increase the fraction of filled pores. Under conditions resembling the physiological state the fraction of filled pores is always greater than 0.9. Because f is close to 1, the flux is approximately equal to N/τ, which means that Q is proportional to (ΔPR_p — const.). Thus changing either ΔP or R_p will have a similar effect on the net efflux of cells. Near the threshold, small changes in ΔP or R_p will produce large fractional changes in the cell flux. Far from the threshold the effect of changing ΔP or R_p will be significant, but the fractional change in the flux will not be as large.

CONCLUSIONS

The goal of this analysis was to identify physical parameters involved in cell egress in the marrow and to evaluate their relative importance to the egress process. Clearly, factors other than those considered in this work are likely to be important in the living marrow. The white blood cell is motile, and it is possible, even likely that motility plays a key role in white cell egress. Chemotaxis may also be an important factor both in the formation of the marrow pores and in the subsequent cell passage. In this paper we have examined the effects of hydrostatic pressure, pore dimensions, and intrinsic cellular resistance to deformation on the egress process and have shown how these quantities could play a role in the regulation of the process. Our ability to critically evaluate the ideas which we have proposed has been hampered by the lack of reliable data for such fundamental quantities as the density of pores in the marrow, the surface area of the endothelial barrier, the distribution of pore diameters, the magnitude of the pressure gradients within the marrow and the egress times for different types of cells. More data both on the architecture of the marrow spaces and the intrinsic properties of the migrating cells will be needed to improve our understanding of the regulation and control of this vital process.

ACKNOWLEDGMENTS

The authors thank Dr. Marshall A. Lichtman for his cooperation in supplying electron micrographs for our use and for his invaluable advice and expertise regarding hematological aspects of the marrow.

This work was supported in part under P.H.S. grant Nos. HL18208 and HL26485. Additional support was obtained under contract with the U.S. Department of Energy at the University of Rochester Department of Radiation Biology and Biophysics. This paper has been assigned report No. UR-3490-2303.

REFERENCES

Cokelet GR (1981): Dynamics of erythrocyte motion in filtration tests and in vivo flow. Scand J Clin Lab Invest 41 Suppl 156:77–82.

Dabrowski A, Szygula Z, Miszta H (1981): Do changes in bone marrow pressure contribute to the egress of cells from bone marrow. Acta Physiol Pol 32:729–736.

Donahue DM, Reiff RH, Hanson ML, Betson Y, Finch CA (1958): Quantitative measurement of the erythrocytic and granulocytic cells of the marrow and blood. J Clin Invest 37:1571–1576.

Evans EA (1973): New membrane concept applied to the analysis of fluid shear- and micropipette-deformed red blood cells. Biophys J 13:941–954.

Evans EA, Hochmuth RM (1976): Membrane viscoelasticity. Biophys J 16:1–12.

Evans EA, Skalak R (1979): Mechanics and thermodynamics of biomembranes. CRC Crit Rev Bioeng 3:181–330.

Evans EA, Waugh R, Melnik L (1976): Elastic area compressibility modulus of red cell membrane. Biophys J 16:585–595.

Happel J, Brenner H (1965): Low Reynolds Number Hydrodynamics. Englewood Cliffs, NJ: Prentice-Hall, p 153.

Hochmuth RM, Worthy PR, Evans EA (1979): Red cell extensional recovery and the determination of membrane viscosity. Biophys J 26:101–114.

Klebanoff SJ, Clark RA (1978): The Neutrophil: Function and Clinical Disorders. New York: North Holland Publishing Co., pp 74–77.

Lichtman MA, Chamberlain JK, Santillo PA (1978): Factors thought to contribute to the regulation of egress of cells from marrow. In Silber R, LoBue J, Gordon SA (eds): "The Year in Hematology." New York: Plenum Press, pp 243–279.

Michelsen K (1967): Pressure relationships in the bone marrow vascular bed. Acta Physiol Scand 71:16–29.

Schmid-Schönbein G, Sung K-LP, Tozeren H (1981): Passive mechanical properties of human leukocytes. Biophys J 36:243–256.

Skalak R, Impelluso T, Schmalzer EA, Chien S (1983): Theoretical modeling of filtration of blood cell suspensions. Biorheology 20:41–56.

Waugh RE (1982): Surface viscosity measurements from large bilayer vesicle tether formation. I. Analysis. Biophys J 38:19–27.

White Cell Mechanics: Basic Science and
Clinical Aspects, pages 237–254
© 1984 Alan R. Liss, Inc., 150 Fifth Avenue, New York, NY 10011

Adhesion and Locomotion of Neutrophil Leucocytes on 2-D Substrata and in 3-D Matrices

J.M. Lackie and P.C. Wilkinson

*Department of Cell Biology, Glasgow University, Glasgow, G12 8QQ.
(J.M.L.), and Department of Bacteriology and Immunology, Glasgow
University, Glasgow, G11 6NT (P.C.W.), Scotland*

INTRODUCTION

It is a commonly held belief that there should be some relationship between the rate of locomotion of a cell and the strength of the cell-substratum adhesion. The prevalence of this belief does not, however, constitute evidence for it and we will begin by reexamining the basis for the expectation. The general thesis is simple: The cell must gain traction to move; adhesion to the substratum gives traction: If the adhesion is too strong or too weak then movement will be affected. Parts of the argument are undoubtedly correct: Newton's third law tells us that forward movement of the cell must be matched by a rearward reaction upon the substratum or matrix. The problem is in the means by which motive force generated by the cell is translated into displacement. Some of the difficulties may derive from confusion between anchorage and adhesion and our inability to transfer our macroscopic experiences of movement to the microscopic level where the relative contributions of inertia, viscous drag, and viscoelasticity are very different. Adhesion is only one way of gaining a purchase and a number of alternative methods should be considered, although not all will necessarily be available to cells. Even in our macroscopic experience adhesion per se is not a prerequisite for reacting, in a Newtonian sense, against the environment. A coat-hook does not adhere to the rail. The child's head stuck between the railings may offer considerable problems in removal—but is not, barring superglue, stuck in an adhesive sense.

Locomotion of a cell depends upon gaining traction, but different environments may offer different possibilities in terms of the methods available for gaining traction, and the behaviour in one circumstance may not reflect the response in another. Thus, as will be developed more fully below, the requirements for movement over planar surfaces are probably very different from those for movement through deformable matrices of complex character. Movement over a two-dimensional (2-D) substratum probably requires an adhesive interaction; movement in a collagen gel may not—and may even be hindered by adhesive drag.

One way of subdividing the topic is to separate circumstances where adhesion is important from those in which it appears not to be: I shall therefore discuss movement over 2-D substrata and through three-dimensional (3-D) matrices in separate sections. For leucocytes this can be justified in an alternative way—the two sections relate to movement over the luminal surface of the endothelial cell prior to diapedesis and subsequent movement between the endothelial cells and through the connective tissue matrix.

MOVEMENT OVER PLANAR SUBSTRATA

Adhesion

Provided we are considering a smooth substratum, one on which curvatures have radii considerably greater than the dimensions of the cell so that there are no grooves or protrusions to push against or hook around, adhesion would seem to be the only means of gaining purchase. In order to move, the cell must form adhesions with the substratum, use these adhesion anchorage points to transmit the forces generated by the motile machinery and then relinquish these adhesions in favour of an appropriately located distal set of contacts.

The adhesions which are formed must be transient if irreversible anchorage is not to occur. It is therefore pertinent to ask what, in this particular context, we mean by an alteration in the cell-substratum adhesion. If the adhesion is transient then its absolute value is essentially irrelevant provided it exceeds a certain minimum value—that minimum required to transmit all or part of the "action" of the machinery to the substratum. Above this minimum value of adhesive interaction, variation will only be significant if the strength of the adhesion is, in some way, related to the duration of the adhesion. If adhesions persist then the cell will have difficulty in moving on—but a small change in adhesion lifespan might be accommodated by adopting (perforce) an elongated form, becoming a tracked rather than a wheeled vehicle. One might predict therefore that the more transient the adhesion the more rounded will be the moving cell—until the adhesion decays faster than the machinery can work and purchase is lost.

If this argument is correct then we should be measuring not the *strength* of the adhesive interaction but its *duration* (and this will apply to studies of adhesiveness in general).

Adhesion assays in which cells have the opportunity to spread are, on the above argument, assays in which longlasting adhesions will be regarded as strong adhesions because a flattened cell is harder to distract with shearing forces. Thus we have often tended to regard surfaces on which cells become very flattened as very adhesive surfaces and have argued from this that the strength of the adhesion leads to irreversible anchorage whereas the problem may well have been of adhesion life span. Because rounded cells are easily washed away in rinsing a coverslip, we have considered substrata to which only brief adhesions occur as "nonadhesive" or "poorly adhesive." The confusion between longlasting and strong adhesions also has implications for aggregation assays: Transient adhesions are less likely to allow mutual spreading in a doublet, and the opportunity to recruit other adhesion sites to stabilize the aggregate will be lost and so the extent of aggregation will be reduced. The parallel plate flow chamber assay developed by Doroszewski [1980] and used recently by ourselves [Forrester et al, 1983] may partially avoid the problem since a more-or-less instantaneous interaction is required to render the cell stationary—certainly the time factor in the assay is much reduced.

We are hindered in approaching the problem of determining the life span of adhesions by our ignorance of the nature of the adhesive interaction. Although prejudice might favour receptor-ligand interactions it is by no means clear that this is invariably or necessarily the case.

Possible Sources of Transience in Adhesions

The absence of hard evidence about adhesion mechanism does not prevent us from speculating about ways in which transience of adhesions might be achieved. The following list is probably incomplete, but will serve as a starting point for discussion:

(A) *Receptor-ligand interactions*
(1) Protease or glycosidase attack on either the receptor or the ligand.
(2) Conformational change in the receptor leading to release of ligand
   i) occurring as a function of receptor age,
   ii) facilitated by, or a consequence of, receptor occupancy.
(3) Receptor degradation following upon, or resulting in, conformational change.

(B) *Physicochemical interactions*
      (1) Increased electrostatic repulsion forces.
      (2) Decreased attractive forces.

(A)(1) Receptor-ligand: Enzymic cleavage. An important factor will obviously be the nature of the receptor involved: For neutrophil leucocytes we can identify receptors for some interactions—the Fc and C3b receptors—which are of obvious physiological relevance. In these cases both the participants are potentially protease sensitive whereas in lectin-carbohydrate interactions only the lectin will be susceptible, although since the carbohydrate moieties of interest are likely to be protein linked they may still be susceptible. Most probably the receptor-ligand complex will be protease sensitive—yet locomotion is unaffected and adhesion is actually decreased by antiproteases, nor does the addition of exogenous proteases have much effect [Forrester et al, 1983]. The argument is weak, however, because the receptor-ligand complex is probably rather inaccessible and an intrinsic membrane protease, operating as it must immediately next to the complex, will not easily be inhibited.

(A)(2) Conformational change. Conformational change in receptors is well documented in other contexts and offers a wide range of possibilities. The receptor must link in some way with the motile machinery of the cell, probably with actin microfilaments of the cortical meshwork, as well as with the extracellular substratum-bound ligand. The requirements for interaction on the cytoplasmic face of the plasma membrane means that control systems might operate intracellularly rather than extracellularly. Binding of F-actin to the receptor might facilitate ligand binding—or vice versa; release of the F-actin might occur once the motile machinery has been activated and local cytoplasmic calcium ion concentrations are altered. It is all too easy to visualize schemes by which events within the cell might, through conformational change in the receptor, influence the properties of the outer surface of the cell, and somewhat more difficult to devise good experimental approaches. Further difficulties arise because receptor replacement and receptor recycling are both likely to require metabolic activity by the cell and straightforward inhibition of the conformational change in the receptor will be difficult to achieve.

(A)(3) Receptor degradation following conformational change. The idea that proteolytic degradation of the receptor on the cytoplasmic face of the plasma membrane might lead to, or be facilitated by, the extracellular binding of ligand is, of course, only one possibility among many, but one which might be worth investigating. In studies on erythrocyte aging, one current model requires proteolytic cleavage, possibly on the cytoplasmic face of the membrane, of a transmembrane glycoprotein (Band 3) which exposes a previously masked antigenic site on the outer surface [Kay et al, 1982]. The details of this are irrelevant to the present argument but the idea of proteolytic cleavage on the cytoplasmic face leading to breakdown of the adhesion is of some interest since a membrane-associated protease with the appropriate orientation has been reported from erythrocytes [Tarone et al, 1979] as well as cytoplasmic proteases [Murakami et al, 1981]. The direct approach to searching for cytoplasmic or membrane-bound proteases, the preparation of plasma membranes or cytoplasmic fractions is, unfortunately, unlikely to be successful in a cell such as a neutrophil, which has massive lysosomal protease depots. If, however, internal proteases are involved in the degradation of adhesions, studies with externally applied proteases and antiproteases might well have produced the results which we reported [Forrester et al, 1983].

Physicochemical interactions. Whilst these cannot be the whole basis of adhesion with cells which do have specific receptors such as those for Fc and C3b, there is no good reason for neglecting the possibility that the interactions with some surfaces depend upon the balance of electrostatic, electrodynamic, and hydrophobic forces. Classically this adhesion hypothesis is based by analogy upon colloid flocculation theories (particularly the DLVO model) and transience of the interaction would not be anticipated. We are, however, dealing with a rather more complex system in which it is not impossible to visualize alterations in crucial parameters as a function of time. Thus by altering the permeability of the plasma membrane the local ionic composition of the gap between cell and substratum might be changed, thereby altering the extent of charge shielding or, indeed, the dielectric properties of the intervening medium. Similarly, by locally altering the protein or glycoprotein concentration in the membrane—by clustering of charged moieties, for example—the balance of attraction and repulsion might shift. Recent work with fibroblasts adhering and moving on artificial substrata of poly-L-lysine or poly-L-histidine [Sugimoto, 1981] supports the idea that physicochemical forces might be important: These particular substrata are interesting in that local pH change would influence the charge on the substratum (the pK of poly-L-histidine is well within physiological range) and alter the electrostatic repulsion.

In the section above I hope I have shown that we have no difficulty in introducing transience into hypothetical adhesions. Whether any of these methods are actually used by cells remains open. Until our assays of adhesion are improved considerably we will remain ignorant of the relative contributions of strength and permanence in adhesion interactions.

Problems With Adhesion Assays

Another problem arises in our adhesion-locomotion studies, which derives from the assay procedures. Our best measurements of locomotion come from time-lapse filming of individual cells (see below), but our adhesion assays are done on populations of cells. Were these populations totally homogeneous this would not matter, but we have good reason to suppose that there will be heterogeneity in adhesion as there is in receptor expression [Klempner and Gallin, 1978; Seligman et al, 1981]. Consider, for example, two substrata of different adhesiveness, and suppose that our neutrophil preparation has "sticky" and "nonsticky" groups or subsets: The sticky cells on the more adhesive surface will form a strong cell-substratum adhesion and may therefore be immobilised through excessive anchorage; the less sticky cells on the less adhesive substratum may be immobile through lack of traction: Different cells from the population will move on the two substrata. If the subsets are of different size in different preparations of cells then the *population* adhesiveness may vary considerably, but those cells which are moving are those which have the appropriate adhesion for the particular test substratum. The proportion of the population which moves may vary depending upon the relative size of each adhesion subset, but if only the cells with the appropriate adhesiveness are moving then the speeds of locomoting cells might well be the same, and this seems to be borne out by experimental evidence. This argument, together with the experimental data, has been presented in more detail elsewhere [Lackie and Brown, 1982, 1983] and will not be reiterated here, although some of the results in a later section of this paper are in accord with the hypothesis.

Thus two major problems exist in measuring adhesion—one the question of duration versus strength discussed in an earlier section, the other that measurements of adhesion are generally based upon populations rather than individuals. In an attempt to evade some of these problems we have recently studied the movement of neutrophils on two spatially adjacent substrata of different properties and looked in detail at the locomotor behaviour of individual cells which cross the boundary. These experimental observations are described in a later section.

QUANTIFYING LOCOMOTION

Not only are there difficulties in quantifying adhesion, but obtaining a good measure of locomotion poses problems. Fortunately these problems are not quite so severe and the measurement of locomotion is more satisfactory than adhesion measurement. For our purposes, as will be discussed later, direct observation of individual cells is the ideal and time-lapse filming provides a good approximation [for review of locomotion assays see Wilkinson et al, 1982].

This leaves the problem of analysing tracks, or rather approximations to tracks—the linked displacements of the cell centre at fixed time intervals. We have chosen to use the method recently described by Dunn [1983], and because of its novelty and because we feel this is an extremely valuable approach, it is described here in some detail.

The film is projected onto paper and the cell centre marked at ten-frame intervals (every 40 seconds real time); the linked dots are an estimate of the real track, a poorer approximation than marking every five-frames, a better approximation than every 20 frames (Fig. 1). Ideally we want the instantaneous velocity—the displacement/unit time, in the limit as the time between position marking is decreased. We therefore take the x-y coordinates of each dot on the track using a digitizing tablet linked to a microcomputer [Lackie and Burns, 1983] and calculate the root mean square displacement between sequential positions (single steps), between alternate positions (double steps), and so on. For reasons both theoretical [see Dunn, 1983] and empirical, we have chosen to take one-, two-, three-, four-, and five-fold steps, fitting the maximum number of steps into each track (20–30 positions on average). Nonoverlapping steps are arbitrary in that the starting position of the cell in the film sequence has no particular significance although our choice of overlapping steps does reduce the independence of the measurements. We then fit, using a least squares method, the best line for the reciprocal of root mean square displacement against the reciprocal of time (Fig. 2). The gradient of this line is an

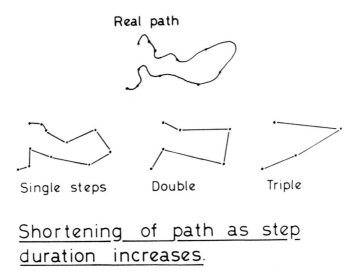

Real path

Single steps   Double   Triple

Shortening of path as step duration increases.

Fig. 1. The "real" track of a cell is shown, together with approximations to the track, taking single, double, and triple "steps" for purposes of illustration. A step is the line between the positions of the cell centre at fixed time intervals. The track length progressively shortens as longer steps are taken, and the speed will therefore be underestimated. Clearly "turning angles" are meaningless, especially when longer steps are used.

estimate of speed (S) and the intercept on the x-axis an estimate of the persistence time (P)—the higher the persistence the straighter the path. From these two parameters we can calculate a "diffusion coefficient" (R) where

$$R = 2 \cdot S^2 \cdot P \tag{1}$$

As a means of estimating the error in our measurement of mean S and P for a group of cells we have used the "Jackknife" method of Mosteller and Tukey [1977], calculating pseudovalues of the mean line-fit parameters by omission in turn of each one of the set of tracks and handling the pseudovalues as a normally distributed data set using parametric statistics.

Our confidence in this approach to analysis of locomotion is greatly strengthened by the similarity in the values obtained from different films of the same cells and by having a set of tracks analysed by Dr. G. A. Dunn

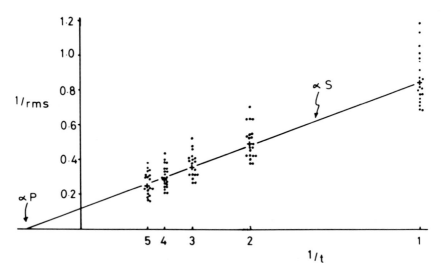

Fig. 2. The data for the individual cells of sequence 1 of Table I moving on a substratum of bovine serum albumin, on the double reciprocal plot of root mean square (rms) displacement against time. The times on the x-axis are for single, double, triple, quadruple, and pentuple steps, the maximum number of each step type being fitted into each track. Individual points are shown as dots, the computed mean of rms displacement as crosses. The gradient of the line is 1/S, where S is the speed; the intercept on the x-axis is $-1/(6P)$, where P is the persistence time.

using a more sophisticated computing technique, in which the expression describing movement

$$<d^2> = Rt - RP[1 - e^{-t/P}] \qquad (2)$$

(where R is the diffusion coefficient, t is the time over which displacements (d) are measured, and P is the persistence time) was fitted by a least squares method, minimising on R and P and weighting data by the reciprocal of the variance. The values for speed and persistence derived in this way differed from our values, obtained using the linear approximation method of a double reciprocal plot of root mean square displacement against time, by less than 1%.

LOCOMOTION OF NEUTROPHILS ON DIFFERENT SUBSTRATA

Despite the problems of measuring adhesion our starting question remains, Does the adhesiveness of the substratum (or the duration of cell-substratum adhesive interaction) affect the rate at which a cell will move over that surface? Is there an optimum level of cell-substratum interaction for movement, or is the neutrophil insensitive to variation in the surface over which it is moving?

In view of the arguments developed in earlier sections the simple approach to studying the interaction of adhesion and locomotion, of carrying out standard adhesion and locomotion assays and then trying to interpret the combined results, has to be abandoned. Assays which test single cells both for adhesion and for locomotion are needed but are not available. Interestingly, Keller [1983] finds a good correlation between the area over which the neutrophil makes a close approach to the substratum (judged by interference–reflection microscopy) and the rate of locomotion, as does Sugimoto [1981] for fibroblasts (estimating a "contact-index" using electron microscopy)—yet we would not normally express adhesion in units of area! Indirect approaches such as these or the ones which we reported in earlier papers [Lackie and Brown, 1982, 1983] may be the only way of gaining insights into the adhesion-locomotion interaction, and another indirect approach is described below.

We have chosen to look at two substrata upon which the nature of the adhesive interaction differs; both are of some physiological relevance. The simple substratum is of bovine serum albumin (BSA) adsorbed onto a glass coverslip; the more complex is the same substratum which has been further coated with immunoglubulin (IgG) directed against BSA (anti-BSA, aBSA) by the simple expedient of touching the BSA-coated coverslip with a second coverslip on which anti-BSA is adsorbed. This gives us two territories: one with an albumin coat and one coated with albumin-antialbumin complexes with the Fc moieties exposed and with which we hope the Fc receptors of the neutrophils will interact. By filming a field with both territories visible we can look at boundary crossing and its consequences. The two territories cannot be recognised directly but there is a marked difference in cell behaviour with cells considerably more flattened on the immune-complex coat: We draw an arbitrary straight line as the boundary when analysing the film, and on the basis of the analysis of the behaviour of the cells this seems to be reasonable.

We assume that the interaction with the substratum will be different when Fc receptor-ligand interactions come into play, in that we have substituted a defined high-affinity receptor-mediated interaction for the very low-affinity, nonsaturable interaction with native BSA. That an interaction with the Fc coat does occur is borne out by the observation (Wilkinson, Michl, and Silverstein, unpublished) that neutrophils on such an immune-complex-coated surface fail to form rosettes with IgG-coated erythrocytes, implying that the Fc receptors have been redistributed to the ventral surface of the cell, as is the case with macrophages on such surfaces [Michl et al, 1979].

Results of the Analysis

If we consider those cells which remain within a single territory throughout the film sequence we find that the rate of movement is distinctly lower on the immune-complex-coated substratum although, surprisingly, the persistence time is unaltered (Table I). Since the "diffusion rate" (R) of the population depends upon the square of the speed, and only linearly upon the persistence time, the effect of this speed change in enhancing accumulation on the immune complex territory is more dramatic than might at first appear. The constancy of the persistence time probably means that the turning behaviour of the cells is controlled independently of the mechanism of forward movement which would account for the lack of correlation between speed and "turning" angle (the angle between successive steps as though straight lines between the positions represented the true path) [Allan and Wilkinson, 1978; Lackie and Burns, 1983]. This would also be consistent with the observation that microtubule-associated systems seem to be associated with polarity and orientation [Allan and Wilkinson, 1978; Englander and Malech, 1981] whereas microfilament systems probably constitute the mechanism for forward movement. It should perhaps be pointed out that the estimates we give here for speed take no account of the stopping behaviour of the cell; we are measuring only travelling time even though (as will be seen in Table II) the cells are stationary for a greater proportion of the time on the immune complex surface.

TABLE I. The Speed (S) and the Persistence Time (P) of Human Blood Neutrophils Filmed on Control and Immune Complex-Coated Substrata[a]

Sequence	No. of cells	S (μm/min)	P (sec)
(1) BSA	20	19.9 ± 0.87	42 ± 6
aBSA/BSA	20	13.9 ± 0.71 (70%)	42 ± 8
(2) BSA	23	17.6 ± 0.8	40 ± 6
aBSA/BSA	17	14.3 ± 0.8 (81%)	35 ± 7
(3) BSA	25	14.0 ± 0.6	35 ± 5
aBSA/BSA	17	11.1 ± 0.6 (79%)	39 ± 6
(4) BSA + serum (a)	11	9.8 ± 0.4	59 ± 8
BSA + serum (b)	13	9.4 ± 0.4 (96%)	73 ± 9
aBSA/BSA + serum	12	6.2 ± 0.3 (63%)	78 ± 8

[a]The values given for speed (S) and persistence (P) are the "Jackknife" mean values ± SEM calculated as described in the text. Sequences 1 and 2 were from different films but on the same cell preparation. Sequences 3 and 4 were both with different cell populations; the values from the separate preparations should not be compared directly. In sequence 4 the cells were moving randomly in that part of the field furthest from the immune complexes (4a), whereas nearer the boundary (4b) there was a distinct population displacement (chemotaxis) toward the immune-complex territory. There were 2.8×10^6 molecules of BSA (bovine serum albumin)/100 μm^2 and 2.2×10^6 molecules of IgG/100 μm^2, giving an antibody/antigen ratio of 1/1.25 (using ^{125}I-labelled proteins and correcting for nonspecific adsorption of IgG with a control substratum of ovalbumin). The serum used in sequence 4 was fresh human serum at a final concentration of 20% (v/v).

TABLE II. The Stopping Behaviour of Cells Filmed as Described in Table I

Sequence	% of time stationary[a]
(1) BSA	10
aBSA/BSA	22
(2) BSA	12
aBSA/BSA	25
(3) BSA	15
aBSA/BSA	43
(4) BSA (a)	20
BSA (b)	26
aBSA/BSA	28

[a]A cell was defined as stationary if it did not make a net displacement of at least one cell radius during a ten-frame (40-sec) interval. Some of these cells may have been moving very slowly during such a period, but since the displacement could not reliably be estimated they were considered to be static: These figures may therefore be an overestimate. Cells which did not displace through the whole film sequence have been excluded.

When, however, we look at the behaviour of individual cells as they move from one territory to another we find no significant diminution in the rate of movement when they are on immune complexes (Tables III, IV) and some cells actually accelerate. As with the single-territory populations there is an increase in stopping behaviour which we have ignored in computing speeds.

TABLE III. The Speed of Movement (S) of Individual Cells Before and After Crossing the Boundary of Immune Complex Deposition[a]

S µm/min on BSA	No. steps	S µm/min on aBSA/BSA	No. steps
9.86	15	8.49	12
13.04	12	9.66	19
11.72	7	10.37	25
13.08	17	13.46	10
10.68	10	11.28	10
10.71	14	7.43	13
16.68	19	11.15	12
11.96	13	10.22	19
11.12	8	12.65	14
17.96	11	12.42	17
13.73	14	9.02	9
9.77	13	11.25	8
Mean 12.52 ± 2.58		10.62 ± 1.78	

[a]The mean values for speed of movement are not significantly different—they are shown together with their standard deviations. Steps were of 40-second duration in this film sequence, which has been taken at random to illustrate the variation in speed and response to the boundary. A summary of the behaviour in a larger set of cross-over events is shown in Table IV.

TABLE IV. A Summary of the Behaviour of Cells Crossing From the BSA-Coated Substratum Onto the Immune Complex-Coated Substratum and vice versa

Cells crossing from BSA to aBSA/BSA		
Mean speed (±SD)		
on BSA	13.68 ± 4.29 µm/min	45 cells; 12.8 steps on average
on aBSA/BSA	12.15 ± 4.53 µm/min	45 cells: 9.3 steps on average
Paired-sample t-test: t = 0.055 df 44: not significant		
16 cells accelerate } not significant 29 cells decelerate		

Cells crossing from aBSA/BSA to BSA		
Mean speed ± SD		
on BSA	12.08 ± 1.25 µm/min	9 cells: 8.2 steps on average
on aBSA/BSA	11.60 ± 1.45 µm/min	9 cells: 10.4 steps on average
Paired-sample t-test: t = 0.54 df 8: not significant		
Mann-Whitney U-test: not significant		
5 move faster on BSA } not significant 4 move faster on aBSA/BSA		

An indication that there is indeed a difference between the two territories is the disproportion in numbers which cross in each direction: Many more move from the BSA-coated territory onto the immune complex (aBSA/BSA) than move in the opposite direction (52 move from BSA to aBSA/BSA, 27 move the other way: a significant difference (chi-squared value = 7.9; P = .0049)). The probability of a cell crossing the boundary depends, of course, both upon the rate of movement and the stopping behaviour. Although there *is* a difference in the behaviour on the two surfaces the difference seems largely to be in the extent of the time for which the cells are immobilized and the speed difference is not significantly different, although there is a slight decrease and more cells slow down when they cross onto the immune complex substratum than speed up. If this result is an accurate reflection of the true behaviour then we must attempt to reconcile the absence of speed change in the individual cell with the difference in population speed when we analyse the tracks of cells which remained on one or other of the two territories (Table I). Several factors may be involved. The cells which have just crossed the boundary have not been exposed to the immune complexes for the same length of time as those which have been there for the whole of the film sequence, and receptor redistribution is unlikely to be instantaneous. A proportion of the cells may not express Fc receptors, or express them only at a low level, and these may be selected for in the analysis since we are concentrating on those cells which retain their locomotory capability. Even when we consider the difference in speed between the populations which remain on single territories the change is far from dramatic: Cells do move over the immune complex-coated substratum and at 70–80% of the speed on BSA. This might be for a variety of reasons, an obvious one being that the level of Fc expression on the surface we have used is rather low. When, however, we measure the amount of immune complex deposited on the surface, the calculated number of Fc receptors on an area approximately equal to that of the contact area of a cell is considerably greater than the number of Fc receptors on the whole cell. This does not take account of the possibility that only a proportion of the Fc moieties are available for binding, and a more extensive series of observations would be required to determine the "dose-response" effect of immune-complex on locomotory rate.

These observations are of relevance in terms of accumulation of neutrophils in areas of immune complex deposition but need not necessarily be as simple to interpret as we have pretended. Taking this system as one in which to examine adhesion and locomotory rate we have assumed that the properties of the individual cell remain invariant: But this is unrealistic. We know that Fc receptors become redistributed when the cells are on an immune complex-coated substratum [Wilkinson, Michl, and Silverstein, unpublished], and we know that secretion of lysosomal enzymes is greatly enhanced on such a surface [Henson, 1971] and that enzyme secretion correlates with increased adhesion [Lackie, 1977]. Both these changes would, however, increase the adhesiveness of the cells and might tend to support the general thesis that increased adhesion tends to reduce locomotion and will trap the cells. Whether the strength of the adhesion or the relative permanence is responsible might be considered irrelevant to the functional consequence—which accords with the physiological role of these cells.

LOCOMOTION IN 3-D MATRICES

In order to move over a smooth planar surface it seems likely that adhesion must provide anchorage. This does not, however, seem to be the case when we consider movement in a 3-D matrix of hydrated collagen. If we coat a coverslip with a very thin adsorbed layer of native type I collagen prepared from rat-tail tendons we find that the adhesion is very low, as judged by the rounded morphology of the cells and their tendency to roll or drift in the filming chamber [Brown and Lackie, 1981]. Possibly because of the lack of anchorage, we have been unable to measure cell speeds because we see no active displacement. If we put neutrophils onto the upper surface of a gel of the same collagen then the cells penetrate and move considerable distances [Lackie, 1982; Brown, 1982]. We have interpreted this to mean that adhesion per se is not an essential requirement and that the cells are gaining anchorage by some alternative means. The appearance of moving cells in time-lapse film suggests that they are squeezing through the meshwork and if we make the meshwork denser by increasing the collagen concentration, the movement is slowed down in proportion. Scanning electron micrographs of freeze-dried preparations support the idea that gels made from more concentrated collagen solutions have a smaller meshwork and suggest (Fig. 3) that cells within these hydrated collagen lattices are in contact with relatively few fibres.

Essentially the same behaviour in 3-D matrices is observed with lymphocytes except that they fail to move over most planar surfaces and are not apparently adherent to such surfaces. When placed on collagen or fibrin gels lymphocytes will penetrate as effectively as neutrophils [Haston et al, 1982] although the morphology of the moving cells differs slightly but appreciably from that of neutrophils. Here again anchorage is apparently gained by some mechanism other than adhesion, an impression reinforced by the observation that lymphocytes will anchor on micropore filters with appropriately sized crevices (pore sizes of 3 μm and 8 μm, but not of 0.45 μm). Fanciful analogies have been used to describe the method of anchorage—fist jamming for example—but the basic idea is simple,

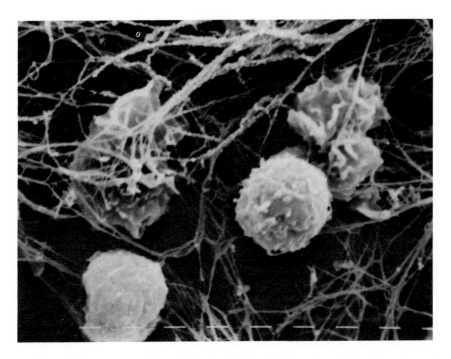

Fig. 3. A scanning electron micrograph showing rabbit neutrophils within a gel of hydrated collagen. The gel was formed from collagen at 1 mg/ml; cells were allowed to invade the gel which was then fixed with glutaraldehyde and critical-point dried. Scale bar = 1 μm.

that by expanding a pseudopod distally it cannot, temporarily at least, be withdrawn from the gap through which it was protruded and can thus serve as a means of reacting upon the environment. Distal expansion of the pseudopod is only one possible method, a change in shape from straight to crooked might serve equally well, but operationally these amount to the same basic mechanism for gaining anchorage.

If this view is correct, that adhesion is not necessarily required for penetration of 3-D matrices, then we could speculate as to the consequences of increasing adhesion of the cells to the fibres which constitute the gel meshwork. One way of enhancing neutrophil aggregation and adhesion, although the mechanism is unknown, is to substitute manganese for magnesium in the buffer solution: if Mn^{2+} is added neutrophils begin to adhere to planar collagen coats and begin to be able to move (Table V). In the 3-D environment this increase in adhesiveness reduces the extent of movement, suggesting that an adhesive interaction with the fibres of the

TABLE V. The Movement of Rabbit Peritoneal Exudate Neutrophils on Collagen-Coated Glass and in 3-D Collagen Matrices With and Without the Addition of Manganese to the Buffered Salts Solution[a]

Medium	Collagen-coated glass			Collagen matrix [leading front in gel (μm) mean ± SD (n)]
	% adhesion (mean ± SE)	% motile cells	% anchored cells	
Mg^{2+}-free	4 ± 1	0	0	164 ± 13 (15)
Control	14 ± 3	0	0	202 ± 16 (15)
10^{-4} M Mn^{2+}	36 ± 9	nd	nd	189 ± 14 (15)
10^{-3} M Mn^{2+}	66 ± 7	74	>20	160 ± 15 (15)*
10^{-2} M Mn^{2+}	110 ± 5	5	>68	57 ± 8 (15)*

[a]The adhesion and locomotion of the cells on the collagen-coated glass was determined in a simple buffered salts solution, modified as shown in the table; gels were reconstituted with the appropriate medium. Cells were considered to be anchored if translocation was attempted but the cell was restrained by posterior adhesions. The figures given may be underestimates as some cells may have been too adhesive even to attempt to move. The sublethal toxicity of manganese cannot be ruled out, but the cells retain their aggregation competence even at high concentrations. [From Brown, 1982, with permission.]
*P (no different from control) < .001.

matrix may actually hinder progress. It is impossible to measure adhesion within these gels but the circumstantial evidence seems quite convincing. Reinterpreting this in terms of the earlier discussion on transience of adhesion, we might suppose that transient anchorage would result from stiffening of the cytoplasm by cross-linking the cortical actin meshwork and by recruiting G-actin into the meshwork, to form the distal protrusion, followed by reversion to a less ordered and more fluid state of cytoplasmic organization in which the pseudopod anchor becomes deformable and ceases to resist tension. An adhesive interaction with the matrix would increase the duration of such anchorages because work would need to be done to withdraw the adherent pseudopod. Control of pseudopod duration, in the nonadhesive situation at least, would therefore be cytoplasmic and very much associated with the control of the motile machinery.

CONCLUDING REMARKS

Neutrophil leucocytes, perhaps more than the other leucocyte classes, must move over and through diverse environments to fulfil their function in vivo: We should not therefore be surprised that they are eminently capable of doing so in vitro. Heterogeneity in the population in respect of adhesiveness or of receptor expression may account to some extent for the capacity of some neutrophils from the population to respond whatever and wherever the challenge. Nevertheless an individual neutrophil must handle diverse problems: movement over planar surfaces and through matrices, and since the properties of connective tissue are likely to vary from place to place, we might also expect that neutrophils would be relatively insensitive to changes in the environment, that the cells should be capable of moving despite local problems. If, for example, the cell is to reach the source of immune complex formation directly then it would be inappropriate to stop when immune complexes are first encountered—at the least a time lag before halting might be expected. We might therefore rationalize the relatively minor changes we observe in rate of movement as we alter environmental conditions as being realistic in terms of the role for which these cells are adapted. This does not, however, preclude alternative explanations!

ACKNOWLEDGMENTS

Dr. A.F. Brown developed the collagen-gel assay during his tenure of an Medical Research Council Studentship, and we are grateful to him for Table V and for Figure 3, as well as for many useful discussions. Lawrence Tetley of the Department of Zoology, University of Glasgow, helped with the scanning electron microscopy, and we are greatly indebted to Dr. G. A. Dunn for help and advice with the analysis of locomotion. Many of our other colleagues have also helped with advice and criticism, in particular Dr. J. V. Forrester.

REFERENCES

Allan RB, Wilkinson PC (1978): A visual analysis of chemotactic and chemokinetic locomotion of human neutrophil leucocytes. Exp Cell Res 111:191–203.

Brown AF (1982) Neutrophil granulocytes: adhesion and locomotion on collagen substrata and in collagen matrices. J Cell Sci 58:455–467.

Brown AF, Lackie JM (1981): Fibronectin and collagen inhibit cell-substratum adhesion of neutrophil granulocytes. Exp Cell Res 136:225–231.

Doroszewski J (1980): Short-term and incomplete cell-substrate adhesion. Symp Br Sco Cell Biol 3:171–198.

Dunn GA (1983): characterising a kinesis response: Time averaged measures of cell speed and directional persistence. In Keller HU, Till GO (eds): "Leukocyte Locomotion and Chemotaxis." Basel: Birkhauser Verlag, pp 14–33.

Englander LL, Malech HL (1981): Abnormal movement of polymorphonuclear neutrophils in Immotile Cilia Syndrome. Cinemicrographic analysis. Exp Cell Res 135:468–472.

Forrester JV, Lackie JM, Brown AF (1983): Neutrophil behaviour in the presence of protease inhibitors. J Cell Sci 59:213–230.

Haston WS, Shields JM, Wilkinson PC (1982): Lymphocyte locomotion and attachment on two-dimensional surfaces and in three-dimensional matrices. J Cell Biol 92:747–752.

Henson PM (1971): The immunologic release of constituents from neutrophil leucocytes. J Immunol 107:1535–1546.

Kay MMB, Sorenson K, Bolton P (1982): Antigenicity, storage and ageing; physiological auto-antibodies to cell-membrane and serum proteins and the senescent cell antigen. Mol Cell Biochem 49:65–85.

Keller H-U (1983): Shape, motility and locomotor responses of neutrophil granulocytes. In Keller HU, Till GO (eds): "Leukocyte Locomotion and Chemotaxis." Basel: Birkhauser Verlag, pp 54–72.

Klempner MS, Gallin JI (1978): Separation and functional characterization of human neutrophil sub-populations. Blood 51:659–669.

Lackie JM (1977): The aggregation of rabbit polymorphonuclear leucocytes (PMNs). Effects of agents which affect the acute inflammatory response and correlation with secretory activity. Inflammation 2:1–15.

Lackie JM (1982): Aspects of the behaviour of neutrophil leucocytes. In Bellairs R, Curtis A, Dunn G (eds): "Cell Behaviour." Cambridge:Cambridge University Press, pp 319–348.

Lackie JM, Brown AF (1982): Substratum adhesion and the movement of neutrophil leucocytes. Microcirc Rev 1:127–133.

Lackie JM, Brown AF (1982): Adhesion and the locomotion of neutrophils on surfaces and in matrices. In Keller HU, Till GO (eds): "Leukocyte Locomotion and Chemotaxis." Basel:Birkhauser Verlag, pp 73–88.

Lackie JM, Burns MD (1983): Leukocyte locomotion: Comparison of random and directed paths using a modified time-lapse analysis. J Immunol Methods 62:109–122.

Michl J, Pieczonska MM, Unkeless JC, Silverstein SC (1979): Effects of immobilized immune complexes on Fc- and complement-receptor function in resident and thioglycol-late-elicited mouse peritoneal macrophages. J Exp Med 150:607–621.

Mosteller F, Tukey JW (1977): Data Analysis and Regression. Reading:Addison-Wesley, ch 8, pp 133–142.

Murakami T, Hatanaka M, Murachi T (1981): The cytosol of human erythrocytes contains a highly Ca^{2+}-sensitive thiol protease (Calpain I) and its specific inhibitor protein (Calpastatin). J Biochem 90:1809–1816.

Seligman B, Chused TM, Gallin JI (1981): Human neutrophil heterogeneity identified using flow microfluorometry to monitor membrane potential. J Clin Invest 68:1125–1131.

Sugimoto Y (1981): Effect on the adhesion and locomotion of mouse fibroblasts by their interacting with differently charged substrates. A quantitative study by ultrastructural method. Exp Cell Res 135:39–45.

Tarone G, Hamasaki N, Fukuda M, Marchesi VT (1979): Proteolytic degradation of human erythrocyte Band 3 by membrane-associated protease activity. J Membr Biol 48:1–12.

Wilkinson PC, Lackie JM, Allan RB (1982): Methods for measuring leucocyte locomotion. In Catsimpoolas N (ed): "Cell Analysis: 1." New York:Plenum Press, pp 145–193.

White Cell Mechanics: Basic Science and
Clinical Aspects, pages 255–268
© 1984 Alan R. Liss, Inc., 150 Fifth Avenue, New York, NY 10011

Lymphocyte Recognition of Lymph Node High Endothelium: Adhesive Interactions Determining Entry Into Lymph Nodes

Judith J. Woodruff and Yee-Hon Chin

Department of Pathology, SUNY Downstate Medical Center, Brooklyn, New York 11203

Large numbers of T cells and B cells enter the blood from thoracic duct lymph, circulate briefly, and then emigrate specifically into peripheral lymphoid organs—lymph nodes, spleen, and intestinal lymphoid tissues. Several hours later, these cells reenter the bloodstream directly from the spleen or indirectly via lymphatics from lymph nodes and the gut. Cells with this life history comprise the recirculating pool of lymphocytes [Gowans, 1957, 1959a,b; Gowans and Knight, 1964; Ford, 1975].

Lymphocyte recirculation delivers T cells and B cells to lymph nodes (LN), intestinal lymphoid tissues (ILT—ie, tonsils, Peyer patches) and spleen. The result is a continuous process of redistribution and reassortment of the cells which comprise these tissues. Recirculation is known to play a central role in the development of normal immune responses in intact animals and it has the potential to exert an important impact on regulation of immune reactivity [Ford, 1975; Ford and Gowans, 1969; Sprent, 1977]. Techniques for control of lymphocyte migration into tissues are of interest since they may provide new means for regulating this reactivity.

The early work on lymphocyte recirculation led to the view that repeated passage of lymphocytes from blood to lymph through LN and Peyer patches (PP) resulted in a complete mixing of the recirculating pool. If correct, this meant that the composition of cells in thoracic duct lymph reflected the composition of cells in effluent lymph in general. This view must now give way to the evidence that there exists separate lymphocyte migration pathways. For example, one subpopulation shows preferential migration through peripheral lymph nodes and another, preferential migration through the gut [Griscelli et al, 1969; Craig and Cebra, 1971; McWilliams et al, 1975; Cahill et al, 1977; Hall et al, 1977].

Interaction between circulating lymphocytes and blood vascular endothelium is crucial for recirculation and reassortment of lymphocytes since the vascular endothelium is the first barrier which migrating lymphocytes must negotiate as they move from the blood into the various lymphoid compartments of the body. In this context, the integrity of the recirculating system can be considered to depend largely on events occurring in LN and ILT because these tissues are the main sites where lymphocytes pass en route from blood to lymph [Gowans and Knight, 1964]. In sheep, where cannulation of lymphatics of individual lymph nodes is possible, it has been shown that 14,000 recirculating lymphocytes pass through the node per second and that this can increase to about 140,000 lymphocytes per second in response to antigen [Cahill et al, 1977]. Even under these conditions specificity is maintained and nonrecirculating lymphocytes, granulocytes, and other free-floating cells in the bloodstream are excluded. There must then by some specialized arrangement for transfer of lymphocytes from blood across the endothelium into lymph nodes.

The first step in entry of blood lymphocytes into LN and ILT is the binding of these cells to venules lined by high endothelium (HEV) [Gowans and Knight, 1964]. The cells enter only by crossing HEV and not by migrating through other vascular endothelium [Gowans and Knight, 1964; Marchessi and Gowans, 1964]. HEV are present in LN, PP, and tonsils, and have been observed in humans as well as in most mammalian species [Miller, 1969]. Typically the specialized endothelium is formed by polygonal cells linked together by discontinuous junctional complexes. Ultrastructurally most high endothelial cells have abundant cytoplasm and contain a prominent Golgi apparatus, vesicles, and fine and clustered ribosomes and mitochondria [Marchessi and Gowans, 1964]. Scanning electron micrographs have shown that the first contact with high endothelium occurs via microvilli on lymphocytes, which are retracted as the cells cross the vessel wall by migrating between, and not through, the endothelial cells [Van Ewijk et al, 1975; Anderson et al, 1976].

Lymphocytes capable of binding to high endothelium were originally identified in rodents by using the adoptive transfer method. It was shown that lymphocytes capable of HEV attachment are in the circulation (blood and lymph) and also in lymph nodes, spleen, and ILT. Both T and B cells have this capability whereas relatively few cells in thymus and bone marrow possess the property [Gutman and Weissman, 1973; Sprent, 1973; Howard, 1972]. The basis of the affinity of lymphocytes for high endothelial cells is not known but it must involve a highly selective mechanism of recognition. The specificity of this interaction and the finding that enzymes which modify surface components of lymphocytes interfere with their accumulation into LN stimulated the idea that this reaction might be mediated by specific membrane molecules [Gesner and Ginsburg, 1964; Woodruff and Gesner, 1968, 1969].

Our approach to this problem has been to use an in vitro method to study cell recognition mechanisms mediating lymphocyte adherence to the high endothelium of lymph nodes. The distinct morphology of high endothelial cells in rat and mouse lymph node sections made it possible to devise a system whereby their interaction with lymphocytes could be studied in vitro. The high endothelium is exposed in such sections and lymphocytes deposited over the tissue can then be assessed for their capacity to bind these cells.

LYMPHOCYTE ADHESION TO HEV IN VITRO AND ITS RELATIONSHIP TO THE MIGRATION OF CELLS INTO LYMPH NODES IN VIVO

Studies in this laboratory have been carried out using rats since the phenomenon of recirculation and much of the information about the mechanism of this process has come from experiments using these animals. In addition our studies have focused on thoracic duct lympho-cytes (TDL) because nearly all lymphocytes emerging in lymph are recirculating cells; they are also almost all small lymphocytes ($\geq 90\%$) and are known to be a mixture of T cells (65%) and B cells (35%).

TDL adhere specifically to high endothelial cells when overlaid onto rat lymph node sections. Binding occurs rapidly at 7°C with maximal levels of adherence found in 15–30 minutes. The results are quantitated by counting the number of "positive" HEV (venules with two or more adherent cells) per section. TDL adhere to at least 80% of HEV with 30–50% showing heavy binding (six or more adherent cells) under optimal conditions, ie, sections are overlaid with 0.2-ml aliquots of lymphocytes at 3×10^6/ml. Further, about 85% or more of the lymphocytes which adhere to the section bind to the high endothelium even though this comprises only about 1–2% of the total area of this tissue. There is essentially no attachment to other endothelia in the section including those lining capillaries, lymphatic channels, and nonspecialized venules in the cortex or medulla.

In vitro HEV reactivity is a property of recirculating lymphocytes and not a trait of lymphoid cells incapable of homing into lymph nodes in vivo. Thus lymph, lymph nodes, and spleen contain substantial numbers of recirculating lymphocytes and comparable levels of HEV binding are observed in sections overlaid with cells from these organs (Table I). In contrast, thymus and bone marrow are deficient in cells with this capability; the extent of HEV binding in sections overlaid with these populations is 10% or less of that found with TDL. In addition T and B cells isolated from lymph enter LN via HEV and in vitro exhibit comparable levels of binding to these vessels. Thus the in vitro system exhibits the dual specificity with regard to vascular endothelium and lymphocyte type which characterizes lymphocyte-lymph node homing interactions that occur under physiological conditions.

The most important conclusions from this work are that (1) high endothelial binding lymphocytes predominate in peripheral lymphoid tissues and are part of the recirculating pool of cells; (2) acquisition of high endothelial binding properties is a differentiation event; (3) binding to lymph node high endothelium is mediated by surface components shared by recirculating T and B cells, ie, components for adhesion are not class specific.

LYMPHOCYTE ADHESION MOLECULES FOR HIGH ENDOTHELIUM

Our main interest in this in vitro system has been to delineate the recognition mechanisms responsible for adhesion. The strategy of our experiments was based on the idea that (1) this function is mediated by specific molecules on lymphocytes and (2) isolated adhesion molecules, when overlaid onto lymph node sections, would retain their capacity to bind to high endothelium. In this event, the high endothelial structures in

TABLE I. Specificity of Lymphocyte-HEV Adherence Assay[a]

Lymphoid population	% positive HEV/section
TDL	87 ± 3
T cells	84 ± 2
B cells	85 ± 3
Lymph node cells	88 ± 7
Spleen cells	86 ± 2
Thymus cells	4 ± 1
Bone marrow cells	2 ± 1

[a]Values are means of four sections ± SE using the overlaid cell population at 3×10^6 cells/ml. T cells ($\leq 5\%$ sIg$^+$) and B cells ($\geq 95\%$ sIg$^+$) were isolated from thoracic duct lymphocytes (TDL) using the nylon wool filtration method. HEV, high endothelial venules.

these sections which are the sites for lymphocyte attachment would be blocked.

The work was therefore begun with an attempt to isolate a component capable of acting as an inhibitor of TDL binding to HEV in vitro. The source selected for isolation of an inhibitor was thoracic duct lymph. Typically, lymph is collected for 18 hours and during this period, $5–10 \times 10^8$ lymphocytes accumulate in the collection flask. If TDL spontaneously shed adhesion molecules, then the possibility existed that they could be recovered directly from lymph. The results demonstrated that lymph does indeed contain such a component. Originally, this component was termed "inhibitory factor" because of its effect on in vitro TDL-HEV binding, and rabbit antisera against the component as antiinhibitory factor antibody [Chin et al, 1980a,b, 1982]. Because of the biologic characteristics of this factor, it is now designated high endothelial binding factor (HEBF) and antibodies against it as anti-HEBF Ig [Chin et al, 1983]. The evidence to be cited suggests that anti-HEBF recognizes high endothelial adhesion molecules on rat peripheral lymphocytes.

Isolation and Properties of Rat Lymph-Derived HEBF

HEBF has been detected by its ability to bind to high endothelial structures in LN sections and block lymphocyte attachment. Initially the factor was isolated from rat lymph by sequential ion exchange and gel filtration and the material then used for immunization of a rabbit. Thereafter anti-HEBF Ig affinity chromatography was utilized for isolation of the factor.

Thoracic duct lymph depleted of cells and chylomicra was fractionated by $(NH_4)_2SO_4$ and material precipitating between 60 and 80% saturation was the source of crude HEBF. The factor does not bind to diethylaminoethyl (DEAE) (1 mM TRIS, pH 8.3) and DEAE cellulose chromatography separates HEBF from the bulk of protein ($\geq 95\%$) in lymph. An additional twofold enrichment is produced by carboxy methyl (CM)-sepharose chromatography with HEBF eluting between 0.05–0.2 M NaCl. On gel filtration (Sephacryl S-200) HEBF emerges in a single protein peak which elutes at an effluent volume similar to aldolase, suggesting that the active component has a molecular weight (m.w.) of about 160,000.

Large-scale purification of HEBF from lymph is now routinely achieved by affinity chromatography using rabbit anti-HEBF $F(ab')_2$ coupled to Sepharose 4B. SDS-PAGE analysis shows that the factor is composed primarily of three polypeptides with m.w. 175,000, 145,000, and 54,000.

HEBF exerts its effect by binding to HEV and blocking attachment sites where lymphocytes would normally bind; it has no effect on functional properties of lymphocytes themselves. Also, binding of the factor to HEV is stable inasmuch as HEV in pretreated sections (25 μg of HEBF in 0.1 ml) which are thoroughly washed with phosphate-buffered saline (PBS) remain blocked and the adherence of overlaid TDL is prevented. In contrast, HEV binding is normal after TDL are treated with this factor (500 μg per 10^7 cells, 30 minutes at 4°C), and then washed before being overlaid onto untreated sections [Chin et al, 1980b].

The factor is trypsin-sensitive, binds to lentil lectin, and is eluted by 0.3 M α-methyl-D-mannoside. It is stable to heat at 70°C but destroyed by treatment at 100°C. Together, the findings indicate that the active component is a glycoprotein [Chin et al, 1980a,b].

HEBF in lymph is antigenically related to determinants that are present on TDL but not on thymocytes. Therefore we consider that the factor may be derived from shed lymphocyte surface structures. Our working hypothesis is that the factor has an affinity for high endothelium; in vitro soluble HEBF blocks sites where lymphocytes attach. In vivo, however, such structures on TDL participate in reactions that result in adherence to high endothelial cells.

Effect of Anti-HEBF Antibody on Lymphocyte Adhesion to High Endothelium

Studies have shown that anti-HEBF antibody blocks TDL binding to HEV (Table II). Thus TDL treated with intact antibody, washed, and overlaid onto LN sections show an 85% reduction in HEV binding

TABLE II. Anti-HEBF Antibody Blocks Lymphocyte Adhesion to HEV of Lymph Nodes[a]

Group		Antibody treatment	% fluorescent positive cells	% positive HEV/section
A	TDL	Anti-HEBF F(ab')$_2$	70	18 ± 7
		Anti-HEBF Fab	ND	7 ± 1
		Anti-thymocyte F(ab')$_2$	95	81 ± 5
		Normal rabbit F(ab')$_2$	3	85 ± 6
B	B-TDL	Anti-HEBF F(ab')$_2$	90	2 ± 1
		Anti-Ig F(ab')$_2$	95	88 ± 5
		Normal rabbit F(ab')$_2$	2	92 ± 7
C	Section	Anti-HEBF IgG	NA	85 ± 3
		Normal rabbit IgG	NA	94 ± 13

[a] For groups A and B the indicated lymphocytes were incubated with antibody (0.3 mg/10^7 cells) at 37°C for 30 minutes, washed once, and overlaid onto untreated sections. In group C, lymph node sections were treated with antibody (3 mg/ml) at 7°C for 30 minutes, washed, and overlaid with untreated TDL. HEBF, high endothelial binding factor.

compared with TDL treated with nonimmune IgG. Antibody does not act by blocking HEV sites because binding is not reduced when LN sections are pretreated with anti-HEBF IgG, washed, and then overlaid with untreated TDL. Treatment of TDL with Fab fragments of this antibody reduces HEV binding by 70–90% (0.2–0.3 mg Fab/10^7 cells), which rules out the possibility that the effect was Fc mediated or produced by cocapping of lymphocyte surface molecules.

Additional evidence for the specificity of the effect of the anti-HEBF antibody was derived from experiments showing that (1) treatment of TDL with either antithymocyte or anti-IgG antibody, and (2) treatment of TDL-derived B cells with anti-Ig does not interfere with the capacity of these cells to bind to HEV. This occurred even though saturating amounts of both antibodies were used; >90% of TDL were labeled by antithymocyte Ig and ~35% of TDL and ~95% of B cells were labeled by anti-Ig antibody. Together, the results suggest that anti-HEBF antibody recognizes lymphocyte surface structures specifically involved in HEV adherence and that sIg and molecules recognized by antithymocyte antibodies are not components of the HEV recognition structures [Chin et al, 1982].

Antibody treatment of TDL also interferes with entry of transferred ^{51}Cr-labeled cells into LN. Thus, treatment of TDL with anti-HEBF F(ab')$_2$ or Fab reduces the levels of radioactivity in cervical and axillary LN by ~70%, in comparison with values obtained in recipients of TDL treated with nonimmune Ig (Fig. 1). However, antibody treatment does not reduce accumulation of radioactivity in PP. In addition, the amount of radioactivity in spleen, liver, lung, or blood is essentially the same in recipients of anti-HEBF and nonimmune Ig-treated TDL. These results indicate that donor cells, unable to enter lymph nodes, were not diverted to any single site.

These effects of anti-HEBF Ig are specific in that recovery of radioactivity in LN, as well as in other organs, is not altered by (1) treatment of radiolabeled TDL with saturating amounts of antithymocyte F(ab')$_2$ (≥90% of the cells bound the antibody as determined by indirect immunofluorescence) and (2) treatment of radiolabeled B cells with saturating amounts of anti-Ig F(ab')$_2$ (≥95% of the cells bound the antibody as determined by indirect immunofluorescence) [Chin et al, 1982].

Our interpretation of these results is that anti-HEBF antibody recognizes lymph surface molecules involved in binding to high endothelium of lymph nodes. Because the antibody does not affect lymphocyte migration into the spleen, it appears that the high endothelial adherence molecules are different from molecules on lymphocytes that interact with vascular endothelium in the spleen.

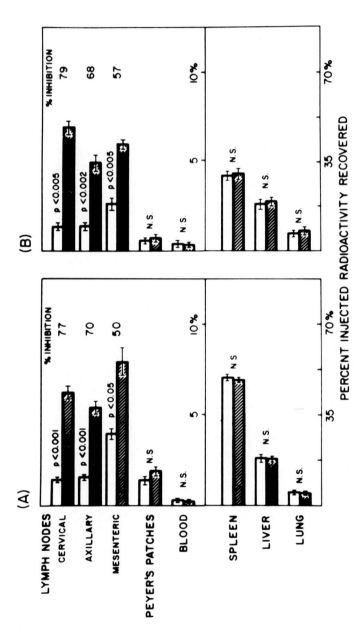

Fig. 1. Lymphocyte entry into lymph nodes blocked by anti-HEBF antibody. Accumulation of radioactivity in organs determined 2 hr after transfer of ^{51}Cr labeled TDL. CPM/recipient of cells treated in vitro with anti-HEBF (▨) or non-immune (☐) (A) F(ab')$_2$: 50,4000 and 49,600, respectively, and (B) Fab: 30,000 and 31,000, respectively. From Chin et al, 1982 with permission.

The evidence that anti-HEBF antibody does not interfere with lymphocyte entry into PP is intriguing, since migration into this tissue also occurs via HEV. This suggests that high endothelial cells differ in LN and in PP at least with respect to the specificity of surface molecules involved in lymphocyte adherence.

Characterization of Lymphocyte-Derived HEBF

The presence of HEBF on different lymphocyte populations was studied by indirect immunofluorescence using rabbit anti-HEBF F(ab')$_2$ (3 mg/10^7 cells for 30 minutes at 4°C) and fluorescein isothiocyanate conjugated (FITC) goat F(ab')$_2$ antirabbit Ig. Antibodies bound to 70% of TDL, 60–65% of spleen and lymph node cells, but to only 2–9% of cells lacking in HEV binding ability (ie, thymus and bone marrow cells; Table III). Therefore, HEBF is found on lymphocytes residing in organs known to contain both HEV binding cells and recirculating lymphocytes.

We then investigated whether HEBF could be isolated directly from lymphocytes. For this purpose, detergent solubilized TDL (500 × 10^6/ml of lysis buffer: 0.05 M TRIS, 0.15 M NaCl, 5 mM ethylene diamine tetraacetic acid (EDTA), 1% Triton, 0.25 M NaDOC, 1 mM phenyl methyl sulfonyl fluoride (PMSF)) were prepared, dialyzed against PBS to remove detergent, and the material then chromatographed on the anti-HEBF F(ab')$_2$ sepharose 4B column. The unbound and eluted (3 M NaSCN) fractions were obtained, dialyzed against PBS, and concentrated to 2 ml. HEBF was detected in the eluted fraction; it caused an 80% reduction in HEV binding and was active to a dilution of 1/16. In contrast, HEBF was not detected in the unbound fraction [Chin et al, 1983].

SDS-PAGE of TDL lysates fractionated by anti-HEBF Ig affinity chromatography revealed that the eluted fraction which contains HEBF is composed predominantly of two major bands with m.w. of approximately

TABLE III. Organ Distribution of Lymphoid Cells Expressing Surface HEBF[a]

Cell source	% labelled cells	
	Anti-HEBF F(ab')$_2$	Nonimmune F(ab')$_2$
Thoracic duct lymph	71	2
LN	62	1
Spleen	65	3
Thymus	2	0
BM	9	1

[a]Cells (2.5 × 10^6) incubated first with the indicated rabbit F(ab')$_2$ and, after washing, with FITC-conjugated affinity-purified goat F(ab')$_2$ antirabbit F(ab')$_2$ under saturating and noncapping conditions; 500 cells counted/sample [from Chin et al, 1983, with permission].

200,000 and 54,000 (unreduced); these are not present in the unbound fraction.

Comparable levels of HEBF were recovered when lysates of TDL, LN, and spleen cells (each at 10^8 cell equivalent/ml) were chromatographed on the anti-HEBF F(ab')$_2$ column, but little if any, was obtained from lysates of thymus and bone marrow cells (Fig. 2). Thus isolation of the factor from lysates was achieved using cells from compartments where HEV binding lymphocytes predominate.

Previous results showed that trypsinized TDL do not adhere to HEV in vitro [Woodruff et al, 1977] and lack the capacity to enter LN in vivo [Woodruff and Gesner, 1969] presumably because the enzyme destroys surface molecules responsible for HEV recognition. It was therefore of interest to determine if HEBF could be recovered from lysates of trypsinized TDL. To study this, TDL were incubated briefly with trypsin, the enzyme was removed, and cell lysates were prepared and chromatographed on the anti-HEBF F(ab')$_2$ column. No HEBF was recovered from these cells, whereas normal amounts of HEBF were recovered from lysates of control cells, ie, TDL treated with premixed trypsin and soybean trypsin inhibitor (mock trypsinized).

These results are of particular importance because they provide the strongest evidence that soluble HEBF isolated from cells is derived from

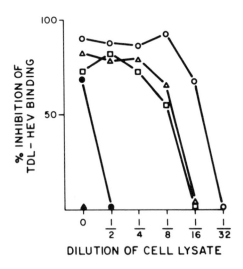

Fig. 2.   Isolation of HEBF from lysates of various lymphoid cells by anti-HEBF Ig affinity chromatography. Each lysate contained 5×10^8 cell equivalent/ml. Lysates were prepared from: TDL, O; LN, △; Spleen, □; Thymus, ●; and bone marrow, ▲. From Chin et al, 1983 with permission.

surface membrane components of lymphocytes. In addition, they provide evidence that destruction of membrane HEBF by trypsin was the event responsible for the failure of such cells to interact with HEV both in vivo [Woodruff and Gesner, 1968] and in vitro [Woodruff et al, 1977].

TDL-derived HEBF exhibits affinity for lentil lectin and is destroyed by treatment with trypsin but not by RNase, properties indicative of a glycoprotein, as previously demonstrated for HEBF isolated from thoracic duct lymph [Chin et al, 1980a,b].

Immunoprecipitation and SDS-PAGE Analysis of Cell Surface HEBF

[125]I-labeled TDL surface proteins recognized by rabbit anti-HEBF Ig have been analyzed by immunoprecipitation and SDS-PAGE autoradiography. Under nonreducing conditions three major bands were observed (m.w. 235,000, 210,000, and 92,000) and under reducing conditions, one major band (m.w. 70,000) and two minor bands (m.w. 92,000 and 30,000) were seen (Fig. 3). The same SDS-PAGE patterns were also observed using ^{125}I-labeled TDL-derived T cells and B cells (Fig. 3), indicating that the composition of HEBF of these populations does not differ and that HEBF is not likely to be a class-specific antigen.

SUMMARY

Our interpretation of these results is that lymphocyte surface HEBF is composed of high endothelial adhesion molecules which mediate cell-cell interactions that result in lymphocyte entry from blood into lymph nodes.

The evidence that anti-HEBF antibody does not interfere with lymphocyte entry into PP is intriguing, since migration into this tissue as well as LN, occurs via HEV. This suggests that high endothelial cells differ in LN and PP at least with respect to the specificity of surface molecules involved in lymphocyte adherence. It could be that separate lymphocyte subpopulations express receptors for these two types of high endothelium. If this explanation is correct, then cells negatively selected using this antibody should be capable of binding to HEV of PP but not HEV of LN. However, if receptors for both high endothelial types are present on the same lymphocyte, then it is unlikely that such selection would yield cells exhibiting tissue specificity. Our observations are consistent with previous findings which suggested that high endothelial cells of mouse LN and PP exhibit differences in lymphocyte binding properties. For example, it has been reported that certain murine lymphomas bind to HEV of either peripheral LN or PP and that, to a limited degree, this preference is a property of most B and T cells [Butcher et al, 1980; Stevens et al, 1982].

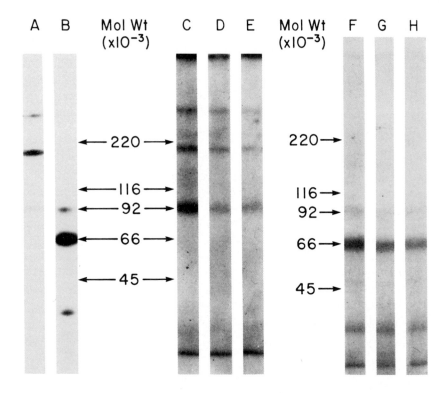

Fig. 3. Analysis of ^{125}I-surface labeled TDL proteins immunoprecipitated with anti-HEBF by SDS-PAGE (5–15% gradient gel) autoradiogram. Lanes A and B: lysates of 10^8 TDL run under non-reducing and reducing conditions, respectively. Lysates of 10^7 TDL, T cells and B cells are in lanes C, D, E (non-reduced), and in lanes F, G, H (reduced) respectively. CPM loaded: lane A, 24,500; B, 26,500; C, 1,300; D, 1,350; E, 1,500; F, 1,600; G, 1,050 and H, 1,400. Lanes A and B were exposed for two days and C-H were exposed for two weeks. From Chin et al, 1983 with permission.

Thymus and bone marrow cells show only low levels of HEV binding and <10% of such cells react with anti-HEBF antibody. During differentiation lymphocytes appear to acquire surface components which mediate high endothelial adhesion and this event is associated with the appearance of surface molecules recognized by anti-HEBF antibody. This suggests that it is the expression of these surface molecules during differentiation which confers high endothelial recognition properties on lymphocytes and that this mechanism plays a role in adhesive interactions leading to entry of both T and B cells into LN.

ACKNOWLEDGMENTS

The authors thank Darien Wilson for expert technical assistance. This work was supported by NIH grant AI 10080.

REFERENCES

Anderson AO, Anderson NO (1976): Lymphocyte emigration from high endothelial venules in rat lymph nodes. Immunology 31:731.

Butcher EC, Scollay R, Weissman IL (1980): Organ specificity of lymphocyte interaction with organ-specific determinants on high endothelial venules. Eur J Immunol 10:556.

Cahill RNP, Poskitt DC, Frost H, Trnka Z (1977): Two distinct pools of recirculating T lymphocytes: Migratory characteristics of nodal and intestinal T lymphocytes. J Exp Med 145:420.

Carey GD, Chin YH, Woodruff JJ (1981): Lymphocyte recognition of lymph node high endothelium. III. Enhancement by a component of thoracic duct lymph. J Immunol 127:976.

Chin YH, Carey GD, Woodruff JJ (1980a): Lymphocyte recognition of lymph node high endothelium. I. Inhibition of in vitro binding by a component of thoracic duct lymph. J Immunol 125:1764.

Chin YH, Carey GD, Woodruff JJ (1980b): Lymphocyte recognition of lymph node high endothelium. II. Characterization of an in vitro inhibitory factor by antibody affinity chromatography. J Immunol 125:1770.

Chin YH, Carey GD, Woodruff JJ (1982): Lymphocyte recognition of lymph node high endothelium. IV. Cell surface structures mediating entry into lymph nodes. J Immunol 129:1911.

Chin YH, Carey GD, Woodruff JJ (1983): Lymphocyte recognition of lymph node high endothelium. V. Isolation of adhesion molecules from lysates of rat lymphocytes. J Immunol 131:1368.

Craig SW, Cebra JJ (1971): Peyer's patches: An enriched source of precursors for IgA producing immunocytes in the rabbit. J Exp Med 134:188.

Ford WL (1975): Lymphocyte migration and immune responses. Prog Allergy 19:1.

Ford WL, Gowans JL (1969): The traffic of lymphocytes. Semin Hematol 6:67.

Gesner BM, Ginsburg V (1964): Effect of glycosidases on the fate of transfused lymphocytes. PNAS 52:750.

Gowans JL (1957): The effect of the continuous re-infusion of lymph and lymphocytes on the output of lymphocytes from the thoracic duct of unanaesthetized rats. Br J Pathol 38:67.

Gowans JL (1959a): The recirculation of lymphocytes from blood to lymph in the rat. J Physiol (Lond) 146:54.

Gowans JL (1959b): Lymphocyte recirculation. Br Med Bull 15:50.

Gowans JL, Knight EJ (1964): The route of recirculation of lymphocytes in the rat. Proc R Soc Lond [Biol] 159:257.

Griscelli C, Vassalli P, McCluskey RT (1969): The distribution of large dividing lymph node cells in syngeneic recipient rats after intravenous injection. J Exp Med 130:1427.

Gutman GA, Weissman IL (1973): Homing properties of thymus independent follicular lymphocytes. Transplantation 16:176.

Hall JG, Hopkins J, Orlans E (1977): Studies on lymphocytes of sheep. III. Destination of lymph-borne immunoblasts in relation to their tissue of origin. Eur J Immunol 7:30.

Howard JC (1972): The life-span and recirculation of marrow-derived small lymphocytes from rat thoracic duct. J Exp Med 135:185.

Marchesi VT, Gowans JL (1964): The migration of lymphocytes through the endothelium of venules in lymph nodes: An electron microscopic study. Proc R Soc Lond [Biol] 159:257.

McWilliams M, Phillips-Quagliata JM, Lamm ME (1975): Characteristics of mesenteric lymph node cells homing to gut-associated lymphoid tissue in syngeneic mice. J Immunol 115:54.

Miller JJ, III (1969): Studies of the phylogeny and ontogeny of the specialized lymphatic venules. Lab Invest 21:284.

Sprent J (1977): Recirculating lymphocytes. In Marchalonis, JJ (ed): "Lymphocyte: Structure and Function. Part 1." New York: Marcel-Dekker, p 43.

Stevens SK, Weissman IL, Butcher EC (1982). Differences in the migration of B and T lymphocytes: Organ-specific localization in vivo and the role of lymphocyte endothelial cell interaction. J Immunol 128:844.

Van Ewijk W, Brons NHC, Rozing J (1975): Scanning electron microscopy of homing and recirculating lymphocyte populations. Cell Immunol 19:245.

Woodruff JJ, Gesner BM (1968): Lymphocyte recirculation altered by trypsin. Science 161:176.

Woodruff JJ, Gesner BM (1969): The effect of neuraminidase on the fate of transfused lymphocytes. J Exp Med 129:551.

Woodruff JJ, Katz IM, Lucas LE, Stamper HB, Jr (1977): An in vitro model of lymphocyte homing. II. Membrane and cytoplasmic events involved in lymphocyte adherence to specialized high-endothelial venules of lymph nodes. J Immunol 119:1603.

V. THE CONTRIBUTION OF WHITE CELLS TO PATHOLOGIC FLOW STATES

White Cell Mechanics: Basic Science and
Clinical Aspects, pages 271–283

Leukocyte Rheology in Cardiac Ischemia

Mark D. Dahlgren, Michael A. Peterson, Robert L. Engler,
and Geert W. Schmid-Schönbein

*Research Service and Cardiology Section, Department of Medicine,
Veterans Administration Medical Center (M.D.D., M.A.P., R.L.E.) and
Department of Applied Mechanics and Engineering Sciences—
Bioengineering (G.W.S.-S.), University of California, San Diego, La Jolla,
California 92093*

INTRODUCTION

Incomplete return of blood flow following periods of ischemia has been observed in the heart [Kloner et al, 1974a,b; Jarmakani et al, 1974], skeletal muscle [Bagge et al, 1980], and in the brain [Yamakawa, 1982]. This has been designated the "no-reflow" phenomenon and in the case of the heart a number of different mechanisms have been suggested which may lead to this condition: loss of volume control in myocardial cells under hypoxia may lead to swelling of myocardial tissue, thus compressing arterioles or capillaries [Flores et al, 1972; Kloner et al, 1974b]. Endothelial cell swelling may lead to narrowing of capillaries [Kloner et al, 1974b] and occlusion of microvessels due to hemorrhage, which has been observed after reperfusion following more than 6 hours of ischemia [Higginson et al, 1982]. Although these mechanisms have been suspected of participating in the "no-reflow" phenomenon they have not been shown directly to cause actual occlusion of vessels in the microcirculation.

Several recent studies have shown that during periods of reduced perfusion pressure of the coronary artery there occurs a gradual increase in the vascular resistance [Frame and Powell, 1976; Guyton et al, 1977; Harris et al, 1981]. This increase in resistance is associated with increased vascular permeability [Harris et al, 1981] and is not reversed after the perfusion pressure is restored to normal values. It can still be detected after administration of vasodilatory agents. Although this phenomenon is concurrent with the no-reflow phenomenon, no explanation has been given for it and no relationship to the no-reflow phenomena has been suggested.

In a recent study we proposed an important role of leukocytes in the no-reflow phenomenon [Engler et al, 1983]. It was shown that (1) the no-reflow phenomenon is due to obstruction at the capillary level; arterioles and venules are still perfused. (2) In the nonperfused capillaries an increased number of leukocytes could be observed which, at reperfusion pressures of 75 mm Hg to 90 mm Hg, could not be removed. These leukocytes thus seem to become trapped in capillaries and they may not readily be removed by hemodynamic forces. This phenomenon has been designated as "leukocyte capillary plugging."

In the following we will discuss evidence that supports the hypothesis of microvascular obstruction by leukocytes, with special reference to the coronary circulation. Similar events may also occur in other organs.

MORPHOLOGY AND RHEOLOGY OF LEUKOCYTES

Among the formed elements of the blood the leukocytes are the largest and stiffest components. In the undeformed state the leukocytes have a spherical shape with numerous membrane folds on their surface. These folds provide the cell with more membrane area than is needed for a smooth sphere of equal volume. Measurements of the membrane area including all folds have been obtained from random electron micrographs using a stereological algorithm [Schmid-Schönbein et al, 1980]. The cells with the largest membrane area among the circulating leukocytes were found to be the monocytes ($436 \ \mu m^2$); the neutrophils (with $182 \ \mu m^2$), eosinophils ($206 \ \mu m^2$), and lymphocytes ($264 \ \mu m^2$) have smaller values. During osmotic swelling of the cell, the membrane areas remain constant within error of measurement. Therefore, the cells swell by unfolding their membrane. The observation of a constant membrane area has been confirmed by the micropipette experiment [Chien et al, 1982]. Similar observations were made on the nuclear membranes, the membrane covering the granules, and the outer membrane of the mitochondria.

Measurements of whole cell volume from random electron micrographs in conjunction with a stereological diameter reconstruction show the leukocytes are larger than erythrocytes. There exist systematic differences among the classes of leukocytes with the monocytes as the largest cells (average volume = $234 \ \mu m^3$; diameter = $7.47 \ \mu m$), then neutrophils (average volume = $182 \ \mu m^3$, diameter = $7.03 \ \mu m$), eosinophils (volume = $206 \ \mu m^3$, diameter = $7.31 \ \mu m$), and the lymphocytes as the smallest cells (volume = $116 \ \mu m^3$, diameter $6.2 \ \mu m$). Each class of cells has a

Gaussian distribution of diameter with standard deviation about 0.5 μm. The diameter values are expressed as that of a smooth sphere with the same volume. The height of the membrane foldings is about 0.25 μm under isotonic conditions.

Leukocytes can be observed in the circulation in a passive or an active state. In the passive state they deform only in response to an external applied stress, such as the plasma fluid stress, or adhesive stress to the endothelium [Schmid-Schönbein et al, 1981] whereas in the active state the cells deform spontaneously by projecting protopods [Schmid-Schönbein and Skalak, 1983]. In the circulation most leukocytes remain in the passive state whereas they convert to the active state at the time of emigration across the endothelium into the interstitium.

The mechanical properties of leukocytes have been tested by the micropipette aspiration experiment [Bagge et al, 1980; Schmid-Schönbein et al, 1981; LaCelle et al, 1982]. These experiments indicate that leukocytes have viscoelastic properties with an incompressible cytoplasm. Thus the cells deform only in shear and for the case of small displacement gradients their rheological properties can be closely approximated by a standard solid model. This model consists of a viscous coefficient (μ) and an elastic coefficient in series (k_2) and in parallel (k_1). The coefficients reflect average material properties of the cell and typical values of the empirical viscoelastic coefficients on human neutrophilic granulocytes were found to be

$$k_1 = 275 \pm 119 \text{ dynes/cm}^2 \qquad (1)$$

$$k_2 = 737 \pm 346 \text{ dynes/cm}^2 \qquad (2)$$

$$\mu = 130 \pm 54 \text{ dynes sec/cm}^2 \text{ (poise).} \qquad (3)$$

The values for eosinophils and monocytes are of the same order of magnitude. The coefficients are dependent on temperature, pH, and osmolality [Sung et al, 1982]. These results show that leukocytes are composed of a much stiffer material than erythrocytes. They exhibit a stiff elastic response to short time deformations and the viscous coefficient is about 2100 times larger than the viscous coefficient of hemoglobin solution (0.06 poise).

Leukocytes have a natural tendency to adhere to the vascular endothelium [Atherton and Born, 1972; Schmid-Schönbein et al, 1975; MacGregor, 1980]. They form a common contact area with the endothelium across which a shear stress and a normal stress can be transmitted. During mild inflammation, as it occurs during exposure of the omentum in a laboratory animal, the magnitude of this stress is relatively large and was estimated, on the average, about 400 dynes/cm^2 with a range between 200 dynes/cm^2 and 1,040 dynes/cm^2. The adhesive stresses may vary during the course of an inflammatory reaction and several biochemical and chemotactic agents appear to modulate the adhesive stress to the endothelium or to other substrates [Garvin, 1961; Atherton and Born, 1972; Fehr and Dahinden, 1979; Wiedeman, 1980; MacGregor, 1980; Hammerschmidt et al, 1981].

In summary, granulocytes are larger than erythrocytes, they are much stiffer cells, and they adhere to the vascular endothelium; these facts suggest a role in the occlusion of single file capillaries.

LEUKOCYTE CAPILLARY PLUGGING IN SKELETAL MUSCLE

In a recent study Bagge et al [1980] investigated the role of leukocytes during perfusion of skeletal muscle (cat tenuissimus muscle) during periods of hemorrhagic shock. They labelled the leukocytes with a fluorescent tag (arcidin orange) thereby achieving a better recognition of these cells during intravital microscopy. Reduction of the perfusion pressure leads to a general reduction of cell velocities in all vessels of the microcirculation. But, in addition, these investigators observed some capillaries with single file of cells which at the same time, showed no motion of cells. Upon careful scanning of capillaries without flow, they found that each contained at least one leukocyte and some contained two or three. In some instances stoppage of flow could be observed upon entry of the leukocyte into the capillary [Bagge and Brånemark, 1977], even under normal perfusion pressures, and frequently leukocytes came to rest at a protruding endothelial cell nucleus inside the capillary. The leukocytes were deformed into cylindrical elongated shapes with a large contact area to the capillary endothelium. When the leukocytes became occasionally dislodged recirculation immediately occurred.

Further studies with the isolated rat hind limb preparation [Braide et al, 1983] showed that bolus infusion of leukocytes in an otherwise leukocyte-free perfusion led to a transient increase in vascular resistance. When this experiment was performed with hypotensive perfusion the vascular resistance did not return to control values. The sustained increase in vascular resistance at low flows was positively correlated to the number of leukocytes remaining in the vasculature. At normal perfusion pressures the sustained increase in vascular resistance was very small irrespective of the number of leukocytes which were infused with the bolus. These results illustrate that leukocyte capillary plugging is critically dependent on the level of the perfusion pressure.

LEUKOCYTE PORE PLUGGING IN MILLIPORE FILTERS

When a whole blood suspension is pumped through 5-μm polycarbonate filters at a constant flow rate a gradual increase of the resistance can be observed. This increase of resistance has recently been shown to be associated with plugging of pores by granulocytes [Chien et al, 1982]. The rate of increase of the resistance is dependent on the number of leukocytes in the perfusate, and nearly two-thirds of the leukocyte cause transient or permanent plugging. Direct electron microscopic identification of leukocytes in the pores show that the majority of plugging cells are neutrophilic granulocytes. Many cells remain attached to the millipore filter on the efflux side. Chien and co-workers draw the conclusion that "these results support the concept that leukocyte plugging may have pathophysiological significance in causing microvascular occlusion."

LEUKOCYTE CAPILLARY PLUGGING IN THE CORONARY CIRCULATION

In ischemic dog myocardium, with occlusion of the left anterior descending coronary artery, accumulation of leukocytes in the ischemic tissue has been demonstrated relative to nonischemic tissue [Engler et al, 1983]. The number of leukocytes per unit length of capillary remaining in capillaries of the ischemic tissue after reperfusion correlated with the percent of capillary length that did not reflow. In these studies the left anterior coronary artery was occluded from 1 to 5 hours and then released. After brief reperfusion, the heart was arrested, excised, and the left main coronary artery cannulated. The entire coronary bed was perfused (ex vivo) with 37° Ringer's lactate at 90 mm Hg to clear all formed elements of blood from the blood vessels capable of reflow under these conditions. Carbon black was then infused to mark all flowing blood vessels. After reperfusion with carbon suspension at normal pressure, the microvessels could be divided in a flowing and nonflowing group from observation of histological section (Fig. 1). All arterioles and venules in ischemic and nonischemic tissue contained carbon, and were thus able to flow. In capillaries there was a strikingly different picture: Capillaries in normal tissue contained uniformly carbon (98%), whereas in ischemic tissue only a variable fraction contained carbon. Capillaries with and without carbon were frequently observed to be side by side. Assuming an average capillary length of 102 μm [Bassingthwaighte et al, 1974], then the no-flow (no carbon) capillaries contained on the average about one leukocyte for every unbranched capillary after 5 hours of ischemia. Capillaries which showed no flow usually had high hematocrit and dispersed platelets; no platelet

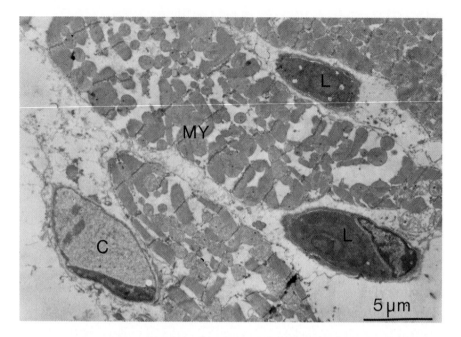

Fig. 1.   Transmission electron microscopic section of single file capillaries in the ischemic myocardium of the dog. The muscle was kept ischemic for 5 hours and then reperfused with carbon suspension at normal pressure and prepared for electron microscopy [Engler et al, 1983]. The cross section shown is a typical view into the ischemic region of the myocardium. The vessel on the left contains carbon particles (C) without cells. The two vessels on the right contain two leukocytes (L) which completely fill the vessel lumen and which could not be removed by the reperfusion pressure. The surrounding myocytes (MY) show edema with rupture of cell membranes.

thrombus could be detected in the biopsies. There was no histological evidence both with high-resolution light microscopy and electron microscopy for true occlusion of capillaries by swollen myocytes or endothelial cells. However, this does not preclude the likelihood of some narrowing of the vessel lumen due to tissue edema during ischemia. The combination of low perfusion pressure, narrowing of vessel lumen due to edema, and increased adhesion stress due to tissue injury may lead to pronounced leukocyte capillary plugging. Our studies relate to permanent obstruction during reperfusion. The aforementioned factors may also play separate roles in initial obstruction, ie, what stops flow, and prevention of reflow, ie, how to start flow again.

In the ischemic tissue leukocyte adhesion to the venous endothelium and occasionally to the arteriolar endothelium could be observed.

Although these leukocytes in such larger vessels may cause an increase in vascular resistance they do not lead to cessation of flow. They may have implications with respect to tissue and endothelium injury.

LEUKOCYTE ACCUMULATION IN THE CORONARY CIRCULATION

To test whether leukocytes accumulate in ischemic myocardium following coronary artery occlusion and reperfusion seven dogs were studied. After left thorectomy the heart was suspended in a pericardial cradle as described previously [Engler et al, 1983]. A short segment of the proximal left anterior descending coronary artery was isolated for occlusion.

A suspension of autologous granulocytes was isolated from a 50-ml blood sample by way of a Ficol Hypaque gradient made of two solutions with specific gravities of 1.135 and 1.077 gm/cm^3. After centrifugation a layer with 80–90% pure granulocytes was removed and again purified with a saline wash and recentrifugation. The cells were labelled with In^{111} by incubation for 30 minutes with 1 mC of In^{111} oxine. The labelled cells were injected into the femoral vein and allowed to circulate for 45 minutes. Then the left anterior descending coronary artery was occluded with a Schwartz clip for periods from 6 minutes to 3 hours. The heart was reperfused for 5 minutes then fibrillated by infusion of KCl, excised, and biopsies from normal and ischemic tissue were taken.

The ischemic tissue was identified by two methods: (1) injection of Evans blue dye into the left anterior descending coronary artery at the site of occlusion, giving a blue coloration to the ischemic tissue. (2) Tissue flow determination by injection of radiolabelled microspheres injected into the left atrium 15 minutes prior to reperfusion. The biopsies were weighed and counted in a gamma well scintillation counter and blood flow calculated by the reference sample technique. Tissues were designated as ischemic if their computed flows were less than 2 SD from the flow in anatomically normal tissue from the circumflex bed.

The results are shown in Table I as In^{111} counts per gram of tissue for endo- and epicardium separately for multiple biopsies in each dog. In every animal, endocardial biopsies from ischemic myocardium contained more labelled granulocytes than normal myocardium with a ratio of counts between 1.6 and 3.0. In four out of five dogs, there was also an increase in granulocyte counts in ischemic epicardium. Two dogs had ventricular fibrillation after 6 and 18 minutes of ischemia; one dog showed a slight but significant leukocyte accumulation in the ischemic tissue.

Previous pathologic studies had noted leukocyte infiltration only after 12–24 hours following the ischemia [Hill and Ward, 1971]. Our studies indicate that there is a significant accumulation of granulocytes after brief

TABLE I. In[111]-Labelled Granulocytes in the Myocardium

Dog No.	Time of ischemia	Endocardium			Epicardium		
		Isch.[a]	Normal	Ratio	Isch.	Normal	Ratio
1	3 hours	3,110 ± 100	1,880 ± 61*	1.65	2,260 ± 171	1,576 ± 81*	1.43
2	3 hours	4,873 ± 342	2,101 ± 246*	2.32	2,619 ± 260	1,929 ± 71*	1.36
3	3 hours	2,054 ± 157	889 ± 55*	2.31	1,121 ± 797	797 ± 20*	1.41
4	3 hours	4,424 ± 511	1,883 ± 115*	2.35	1,514 ± 83	2,416 ± 136	0.63
5	3 hours	10,789 ± 1,181	3,574 ± 507*	3.02	8,977 ± 783	4,377 ± 504*	2.05
6[b]	18 minutes	1,732 ± 52	1,614 ± 30	1.07	1,549 ± 49	1,510 ± 47	1.02
7[b]	6 minutes	1,669 ± 50	1,213 ± 28*	1.37	1,431 ± 96	1,147 ± 26	1.25

[a]Count per minute per gm of tissue, mean ± SEM.
*$P < .001$ ischemic versus not ischemic.
[b]Dogs 6 and 7 had ventricular fibrillation after coronary occlusion, times noted.

ischemia and reperfusion. In the face of leukocyte adhesion to arterioles and venules during the ischemia these studies not only indicate an accumulation of leukocytes in capillaries but may also reflect increased numbers of leukocytes adhering to the endothelium of larger vessels in the coronary microcirculation. Histological investigation of the tissue biopsies also indicated that most leukocytes are still intravascular after 3 hours of ischemia, thereby impairing flow.

LEUKOCYTE MOTION IN CAPILLARIES

As leukocytes enter into single file capillaries they have to be deformed. In heart muscle capillaries this deformation is large. The cells are elongated on the average to about 10 μm and compressed laterally to about 5 μm. In this situation a cell completely fills the lumen of the capillary (Figs. 1, 2) and its membrane has a large common contact area with the endothelium at a gap width of the order of 660 Å. During leukocyte capillary plugging two steps may be distinguished: (1) What leads to initial trapping of the leukocyte and (2) once a leukocyte is trapped in a capillary, what forces are necessary to dislodge it again?

Deformation and trapping of a leukocyte occurs at a tapering segment of the capillary network, such as at the entrance to a capillary [see Bagge and Brånemark, 1977: Fig. 4], or at a point along the capillary with protruding endothelial nucleus. Trapped leukocytes are frequently found at such sites, as indicated by intravital microscopic studies [Bagge et al, 1980].

The problem of deformation of a cell in a narrowing vessel has been studied by Tözeren and Skalak [1979] as the flow of an elastic sphere in a

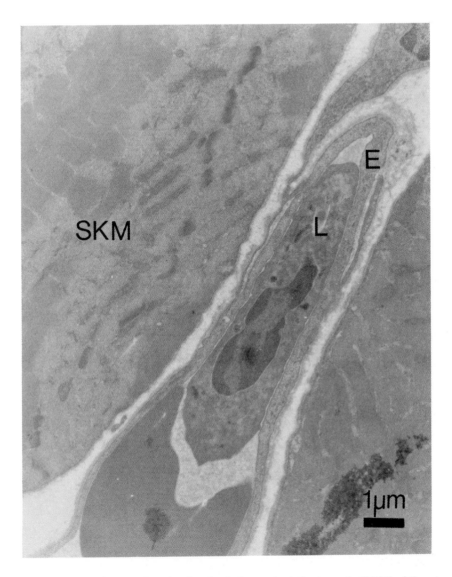

Fig. 2. Longitudinal section of a plugging leukocyte in skeletal muscle (SKM) of the rat spinotrapezius muscle. The leukocyte (L) could not be removed under normal perfusion pressures. There is a close contact between the endothelial cell membrane (E) and the leukocyte membrane.

tapered tube using lubrication theory. As the cell enters into the tapered vessel its velocity decreases significantly after the fluid gap between endothelium and cell has reached a width less than 1 μm. In the fluid gap there exists a compressive pressure which is squeezing the cell laterally and which is opposed by the elastic stress inside the cell to restore its original undeformed spherical configuration. The squeezing pressure is linear proportional to the axial pressure drop pushing the cell along the tube. The cell reduces its velocity significantly as it enters a cone dimension which is within 10% of its cell diameter. Under normal physiological pressure the entry of the leukocyte into a 5-μm tube requires about 1 second [Bagge and Brånemark, 1977], which is significantly longer than that of the erythrocytes (about 1 msec). If the perfusion pressure is reduced, the entry time is increased. The fluid gap width under these conditions is less than the resolution of light microscopy (0.3 μm) and is progressively decreasing with time.

The motion of the cell in such a compressed configuration depends to a large degree on the interaction with the endothelium. In the absence of an adhesive stress to the endothelium the cell will continue to be compressed even in the presence of a reduced pressure drop. This is due to the viscoelastic properties of its protoplasm. Thus, during ischemia the transit time of a leukocyte moving through a capillary would be prolonged, but without adhesion the leukocyte would not stop completely and occlude the lumen.

However, once the endothelial membrane and leukocyte membrane approach each other to within several hundred angstroms an attractive force may be transmitted. This situation is illustrated in Figure 3. Let the shear force per unit area due to this attraction be designated by σ and $\Delta P = P_1 - P_2$ be the hydrostatic pressure drop across the cell; then at equilibrium,

$$\Delta P \pi \left(\frac{d}{2}\right)^2 = \sigma \pi d l, \tag{4}$$

d is the diameter of the capillary, and l is the length of the contact area between leukocyte and endothelium. The term on the *left*-hand side is the resultant axial force acting on the cell due to hydrostatic pressure. The term on the *right*-hand side is the resultant axial force due to adhesion to the endothelium. Upon rearranging terms we find

$$\Delta P = \sigma 4 \frac{l}{d}. \tag{5}$$

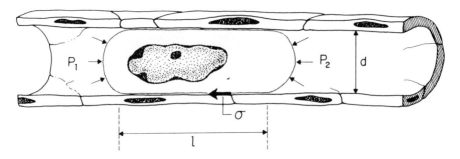

Fig. 3.   Schematic of a leukocyte plugging a straight cylindrical blood vessel. The cell is subject to a pressure drop $P_1 - P_2$ and is attached at the endothelium with an adhesive shear stress σ.

An approximate value of l/d for a leukocyte in a myocardial capillary is about 2 [Engler et al, 1983] so that $\Delta P = 8\sigma$. The cell can only be pushed out of the capillary if ΔP exceeds the maximum shear stress $\sigma = \sigma_{max}$ at which the endothelial and leukocyte membranes separate from each other. Values for σ_{max} have been measured in the rabbit omentum and found to be between 200 and 1,040 dynes/cm^2 [Schmid-Schönbein et al, 1975] (981 dynes/cm^2 = 1 cm H$_2$O) so that a pressure drop of the order of 2 to 8 cm H$_2$O per single leukocyte is required to dislodge it from its plugged position in a capillary. This pressure drop is of the same magnitude as the pressure drop across an unbranched capillary during normal flow. Therefore, within the typical parallel and series arrangement of capillary networks, raising the arterial pressure may not be successful in raising the pressure drop across a single plugged capillary as long as some of its neighboring vessels still flow. In addition, the adhesive shear stress σ_{max} may be a function of the degree of injury to the tissue and may be dependent on the concentration of chemotactic and several other biochemical agents. Under these conditions the pressure drop to dislodge the leukocytes may exceed the values given above.

Furthermore, the contribution of edema with narrowing of the capillary lumen can now be evaluated. If V is the leukocyte volume which in the elongated cylindrical form inside a capillary may be approximated as V = $\pi d^2 l/4$ then, upon substitution into equation 5 one finds

$$\Delta P = \sigma \, \frac{16 \, V}{\pi} \frac{1}{d^3} . \qquad (6)$$

Thus, as the capillary diameter d decreases due to perivascular or endothelial edema, the critical pressure ΔP required to remove a leukocyte increases. For example, a reduction of capillary diameter from 7 μm to 6 μm is associated for constant σ and V with an increase of ΔP by 59% to remove the leukocyte. Further increase of ΔP may occur if the leukocyte swells.

CONCLUSIONS

Current experimental results support the hypothesis that leukocytes play an important role in microcirculatory blood flow, particularly under ischemic conditions. They form a large resistance in single file capillaries and during ischemia may become trapped. Thus an ischemic organ may become a filter for leukocytes. In the myocardial circulation an accumulation of granulocytic leukocytes can be observed early in ischemia (1–3 hours). Thus, the no-reflow phenomenon and the gradual and irreversible increase in hemodynamic resistance may be manifestations of the same phenomenon, ie, leukocyte capillary plugging and microvascular partial obstruction by leukocytes adhering to arteriolar and venular endothelium. This hypothesis finds support by current experiments in our laboratory in the myocardial circulation.

REFERENCES

Atherton A, Born GVR (1972): Quantitative investigation of the adhesiveness of circulatory polymorphonuclear leukocytes to blood vessel walls. J Physiol (Lond) 222:447–474.

Bagge U, Amundson B, Lauritzen C (1980): White blood cell deformability and plugging of skeletal muscle capillaries in hemorrhagic shock. Acta Physiol Scand 180:159–169.

Bagge U, Brånemark P-I (1977): White blood cell rheology. An intravital study in man. Adv Microcirc 7:1–17.

Bagge U, Skalak R, Attefors R (1977): Granulocyte rheology. Experimental studies in an in vitro microflow system. Adv Microcirc 7:29–48.

Bassingthwaighte JB, Yipintsoi T, Harvey RB (1974): Microvasculature of the dog left ventricular myocardium. Microvasc Res 7:229–249.

Braide M, Amundson B, Chien S, Bagge U (1983): Quantitative studies on the influence of leukocytes on the vascular resistance in a skeletal muscle preparation. Microvasc Res (in review).

Chien S, Schmalzer EA, Lee MML, Impelluso T, Skalak R (1983): Role of white blood cells in filtration of blood cell suspensions. Biorheology 20:11–27.

Chien S, Sung KLP, Skalak R, Schmid-Schönbein GW (1982): Viscoelastic properties of leukocytes in passive deformation. In Bagge U, Born GVR, Gaehtgens P, (eds): "White Blood Cells: Morphology and Rheology as Related to Function." Martinus Nijhoff Publishers, The Hague, pp 11–20.

Engler RL, Schmid-Schönbein GW, Pavelec RS (1983): Leukocyte capillary plugging in myocardial ischemia and reperfusion in the dog. Am J Pathol 111:98–111.

Fehr J, Dahinden C (1979): Modulating influence of chemotactic factor-induced cell adhesiveness on granulocyte function. J Clin Invest 64:8–16.

Flores J, DiBona DR, Beck CH, Leaf A (1972): The role of cell swelling in ischemic renal damage and the protective effect of hypertonic solute. J Clin Invest 51:118–126.

Frame LH, Powell KJ (1976): Progressive perfusion impairment during prolonged low flow myocardial ischemia in dogs. Circ Res 39:269–276.

Garvin JE (1961): Factors affecting the adhesiveness of human leukocyte adhesion to glass. Proc Soc Exp Biol Med 141:196–202.

Guyton RA, McClenathan JH, Michaelis LL (1977): Evaluation of regional ischemia distal to a proximal coronary stenosis: Self propagation in ischemia. Am J Cardiol 40:381–392.

Hammerschmidt DE, Harris PD, Wayland JH, Craddock PR, Jacobs HS (1981): Complement-induced granulocyte aggregation in vivo. Am J Pathol 102:146–150.

Harris TR, Overholser KA, Stiles RG (1981): Concurrent increases in resistance and transport after coronary obstruction in dogs. Am J Physiol 240:H262–H273.

Higginson LAJ, White F, Heggtveit HA, Sanders TM, Bloor CM, Covell JW (1982): Determinants of myocardial hemorrhage after coronary reperfusion in the anesthetized dog. Circulation 65:62–69.

Hill JH, Ward PA (1971): The phlogistic role of C3 leukotactic fragments in myocardial infarcts of rats. J Exp Med 133:885–900.

Jarmakani JMM, Cox JL, Graham TC, Hackel DB (1974): Coronary hemodynamics and myocardial perfusion after re-establishing coronary flow in experimental myocardial infarction. Am J Cardiol 33:146a.

Kloner RA, Ganote CE, Jennings RB (1974a): The "no reflow" phenomenon after temporary coronary occlusion in the dog. J Clin Invest 54:1496–1508.

Kloner RA, Ganote CE, Whalen DA, Jenning RB (1974b): Effect of a transient period of ischemia on myocardial cells. Am J Pathol 74:399–422.

LaCelle PL, Bush RW, Smith BD (1982): Viscoelastic properties of normal and pathologic human granulocytes and lymphocytes. In Bagge U, Born GVR, Gaehtgens P (eds): "White Blood Cells: Morphology and Rheology as Related to Function." Martinus Nijhoff Publishers, The Hague, pp 38–45.

MacGregor RR (1980): Granulocyte adherence. Chapter 9. In Glynn LE, Houck JC, Weissmann G (eds): "The Cell Biology of Inflammation. Handbook of Inflammation, 2." Amsterdam: Elsevier/North Holland Publishers, Chapter 9, Amsterdam, pp 267–298.

Schmid-Schönbein GW, Fung YC, Zweifach BW (1975): Vascular endothelium-leukocyte interaction: Sticking shear force in venules. Circ Res 36:173–184.

Schmid-Schönbein GW, Shih YY, Chien S (1980): Morphometry of human leukocytes. Blood 56:866–876.

Schmid-Schönbein GW, Sung KLP, Tözeren H, Skalak R, Chien S (1981): Passive mechanical properties of human leukocytes. Biophys J 36:243–256.

Schmid-Schönbein GW, Skalak R (1983): Leukocyte mechanics during protopod formation. Ann Biomed Eng 11:63.

Sung KLP, Schmid-Schönbein GW, Skalak R, Schuessler GB, Usami S, Chien S (1982): Influence of physicochemical factors on rheology of human neutrophils. Biophys J 39:101–106.

Tözeren H, Skalak R (1979): The flow of closely fitting particles in tapered tubes. Int J Multiphase Flow 5:395–412.

Wiedeman MP (1980): Microscopic observation of small blood vessels in sulfinpyrazone-treated animals. In McGregor M, Mustard JF, Oliver ME, Sherry S (eds): "Cardiovascular Actions of Sulfinpyrozone." Symposia Specialists Inc.

Yamakawa T (1982): Dynamics of white blood cells and red blood cells in microcirculatory networks of the cat brain cortex during hemorrhagic shock: Intravital microscopic study. Microvasc Res 24:218.

White Cell Mechanics: Basic Science and
Clinical Aspects, pages 285–294

Leukocytes and Capillary Perfusion in Shock

Ulf Bagge

*Laboratory of Experimental Biology, Department of Anatomy, University of
Göteborg, Göteborg, Sweden*

The leukocytes are unique among the blood cells, being not only passively transported in the bloodstream but also being capable of leaving the blood circulation by active movements. Observations of the latter phenomenon, ie, the migration of the leukocytes through the vascular endothelium, have made it clear that the leukocytes can undergo large shape changes from the basically spherical shape that they attain when floating freely in the blood.

The average diameters of the undeformed human leukocytes range from 6.2 μm for the lymphocytes to 7.5 μm for the monocytes [Schmid-Schönbein et al, 1981a]. In other words, the leukocyte diameters are almost the same as that of the human erythrocyte, which, on average, is 7.65 μm [Fung et al, 1981]. These blood cell diameters should be put in relation to the diameter of the average systemic capillary, which is about 6 μm [Folkow and Neil, 1971]; in skeletal muscle the average capillary diameter is in fact smaller, about 5 μm according to Eriksson and Myrhage [1972]. From these boundary conditions it is obvious that the deformability of the individual blood cells is of paramount importance for blood perfusion of the capillary bed.

The rheological properties of the individual erythrocyte have been extensively investigated during the past two decades, in practice starting with the in vitro (micropipette) studies performed by Rand and Burton [1964]. Similar in vitro studies on leukocytes were first performed by Adell et al in 1970, followed by studies by Lichtman and Weed [1972], Lichtman [1973], Miller and Myers [1975], Bagge [1975], Bagge et al [1977b], Schmid-Schönbein et al [1981], and Sung et al [1982]. These experiments have shown that the rheological properties of the leukocytes and erythrocytes are very different. This is a fact which was noticed in vivo, however, already by Krogh [1922]. He observed that "the resistance offered to their [leukocytes] passage through vessels with diameters smaller than their own appeared to be greater than that experienced by the erythrocytes." In 1932, Sandison, using the rabbit ear chamber for vital microscopic studies of the microcirculation, made the important observation that even a single leukocyte was capable of completely stopping the circulation, although only temporarily, in a capillary. Nicoll and Webb [1946] studied the microcirculation in the bat's wing and concluded that leukocyte plugging of capillaries was an important cause of intermittent flow in the capillary bed. Nicoll and Webb observed, further, that leukocyte plugging did not only occur at the entrance of the capillaries but also along the course of the capillaries at narrowings caused by the protrusion into the vessel lumen of endothelial cell nuclei. The same authors also stated that the plugging phenomenon occurred under normal flow conditions and that it was not related to an increased tendency of the leukocytes to adhere to the vascular walls. Illig [1957], on the other hand, regarded leukocyte plugging to be the result of sphincter actions, a concept which was refuted by Palmer [1959], who found no evidence of sphincter activities in connection with leukocyte capillary plugging. According to Palmer "it is probable that some degree of leukocyte plugging occurs whenever the diameter of the vessels is appreciably less than that of the leukocyte." Palmer noticed, as we have done repeatedly in our own experiments, that leukocyte plugging can easily be overlooked if too-low optical magnification and resolution are used. Detailed vital microscopic studies of the low-contrast leukocytes certainly require that the optical system (mainly the substage condenser diaphragms) is adjusted for optimal contrast. Another two of Palmer's observations deserve mentioning since they have bearing on the flow behavior of leukocytes in low flow states, to be discussed later. First, Palmer noticed that the duration of leukocyte plugging increased when the general flow was sluggish (low driving pressure) and, second, that at sluggish flows the frequency of leukocyte capillary plugging seemed to increase. This latter phenomenon may be explained by the fact that when the flow is sluggish the erythrocytes tend to form aggregates which will occupy the central stream, distributing the smaller leukocytes toward the peripheral stream [Vejlens, 1938; Palmer, 1967; Nobis et al, 1982]. Being displaced in the periphery the leukocytes should have a greater chance of entering, and possibly blocking, capillary side-branches.

There is some evidence that leukocyte plugging does not cause a complete obstruction of the capillaries but that plasma as well as platelets and erythrocytes in some instances can squeeze their way past the jammed leukocyte [Brånemark and Lindström, 1963]. Due to the flow of plasma past a blocking leukocyte, erythrocytes may enter into the capillary where they will accumulate, densely packed, behind the leukocyte. When the leukocyte moves on the erythrocytes will follow closely behind, forming a leukocyte-erythrocyte train. The velocity of such a train, or of a single leukocyte, in a narrow capillary is lower than that of freely moving erythrocytes [Asano et al, 1973; Schmid-Schönbein et al, 1980b; Nobis and Gaehtgens, 1980] (Fig. 1).

Bagge and Brånemark [1977] performed a high-resolution vital microscopic study of the deformation of leukocytes in the human microcirculation. Their study showed clearly that the leukocytes have viscoelastic properties; the deformation of a leukocyte into a narrow capillary was typically biphasic with an initial rapid but only partial deformation of the cell followed by a comparatively much slower, final adaptation of the cell to the vessel lumen (Fig. 2). It was observed, further, that after passage through a narrow vessel the leukocytes would not retain their spherical shape until after maybe half a minute; the duration of the recovery period was longer the longer the duration of the deformation and/or the more pronounced the deformation. In all cases there was, however, an initial

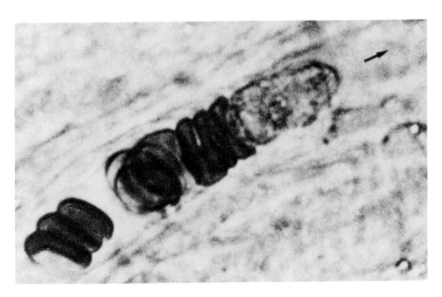

Fig. 1.   Train flow in a human capillary, ie, erythrocytes pile up behind a slowly moving granulocyte. The arrow indicates the flow direction [modified from Branemark, 1971].

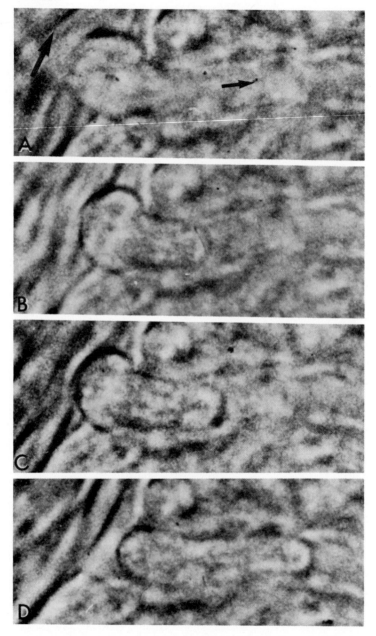

Fig. 2. Leukocyte plugging in man in a 5-μm capillary, illustrating the viscoelastic properties of the leukocytes. The deformation of the cell seen in A required less than 0.06 second; A to B: 0.06 second; B to C: 0.24 second; and C to D: 0.7 second. The arrows indicate the flow directions [modified from Bagge and Brånemark, 1977].

rapid but only partial recovery upon release of the cells. The viscoelastic properties found in the in vivo studies have been verified in several experiments in vitro in different types of glass capillary models [Bagge et al, 1977b; Schmid-Schönbein et al, 1981]. On the basis of these studies a rheological model consisting of a spring in parallel with another spring in series with a viscous element has been proposed to explain the rheological behaviour of the leukocytes. The coefficient of viscosity for neutrophils, monocytes, and eosinophils has been found to be of the order of 130 poise [Schmid-Schönbein et al, 1981], which is almost 2,000 times higher than the viscosity of the hemoglobin solution in the erythrocytes. It should be noted, however, that this value for the viscosity of the leukocytes was obtained at 22°C. More recent experiments at physiological temperature show a value which is about 40% lower [Sung et al, 1982] but, still, the coefficient of viscosity of the leukocytes is more than 1,000 times higher than that of the erythrocyte hemoglobin.

As seen in Figure 2, the degree of deformation required for capillary passage of leukocytes is considerable. This is especially interesting since it is known that the leukocytes deform at constant volume [Bagge and Brånemark, 1977; Bagge et al, 1977b]. However, the deformations are not accompanied by an increase of the surface area since the leukocytes have a wrinkled membrane which provides them with more membrane than is needed to enclose the cell if it were a smooth sphere [Bagge et al, 1977a; Schmid-Schönbein et al, 1980a]. When the leukocytes deform the membrane is simply stretched smooth [Bagge et al, 1977a; see also Bessis, 1973]. If the leukocytes are suspended in a hypo-osmolar medium swelling of the cells will consume the excess membrane area and stiffen the cells [Sung et al, 1982]. Consumption of membrane area may further be one of the explanations of the fact that leukocytes very rarily migrate through the endothelium in narrow capillaries—there may be no membrane left for extension of pseudopods through the vessel wall. For the same reason it seems unlikely that unplugging of a capillary should occur as the result of active, amoeboid movements of the plugging leukocyte.

After acute hemorrhage perfusion pressures are reduced [Mellander and Lewis, 1963] and there are drastic changes in the microvascular flow patterns. In skeletal muscle hemorrhage usually causes a vasoconstriction which may lead to complete cessation of flow for as much as 15–20 minutes, as observed in vital microscopic studies of the cat tenuissimus muscle by Amundson et al [1980a]. According to these authors there is vasoconstriction in central as well as transverse arterioles whereas the terminal arterioles do not constrict. When the flow reappeares the harmonious on-and-off pattern typical of resting skeletal muscle capillary circulation is replaced by an uneven distribution of the flow. The appearance of maldistribution of the muscle microcirculation after hemorrhage has been found also by indirect methods [eg, Appelgren, 1972; Dahlberg, 1979]. A further indication that hemorrhage causes a maldistribution of the muscle blood flow is the finding that it is accompanied by marked local variations in tissue oxygen [Kessler et al, 1976; Lund et al, 1980].

The vital microscopic studies by Amundson et al [1980a] show that on the capillary level the number of perfused vessels is reduced to 30–50% of the normal. Baeckström et al [1971] and Eriksson and Lisander [1972] have proposed erythrocyte and platelet aggregates to be the primary cause of this heterogeneity of the distribution of the blood flow. However, Bagge et al [1980] were not able to correlate any of these phenomena to the disturbed capillary perfusion of the tenuissimus muscle and proposed, instead, that leukocyte plugging of the capillaries could be the reason for the heterogeneous microcirculation. This concept was based on detailed high-resolution vital microscopic scrutiny of all capillaries in which there was no flow. A major problem was, as indicated earlier, the low contrast of the leukocytes which made them very difficult to identify; many leukocytes could only be identified indirectly at first by the peculiar finding that erythrocyte columns resting in the capillaries had a concave rather than an expected convex leading end. As it turned out this pattern was due to the fact that the erythrocytes were pushing at the rounded end of a leukocyte jammed in the capillary lumen. The leukocytes were usually found arrested when endothelial cell nuclei bulged into the capillary lumen or at capillary branch-points (Fig. 3), whereas entrance plugging was observed only in a few cases. The leukocytes did not seem to adhere to the capillary walls but were often seen to oscillate slightly back and forth at the endothelial protrusion. It may seem curious that the leukocytes cannot pass the endothelial cell nuclei which, after all, cause very small (maybe 0.5–1 μm) reductions of the capillary lumen. However, after some period of rest in the capillary the deformed leukocyte will relax, ie, tend to retake its spherical shape, thus pressing itself against the vascular wall. Further, it must be realized that the extra deformation needed may be considerable. Consider the case where an 8-μm sphere is transformed into a cylinder (see Fig. 2) in a capillary with a diameter of 5 μm: The length of the cylinder will then be (if capped with hemispherical ends) about 15 μm. The passage of the cell through a 4-μm-wide section of the capillary will require an additional elongation of the cell of about 7 μm and an additional "increase" of the membrane area of about 20%.

There is experimental evidence that hemorrhage causes leukocyte plugging not only in skeletal muscle but also in the pulmonary microcirculation. Thus, Wilson [1972] has found in histological and electron microscopic examinations of the lungs of dogs a 65% increase in the

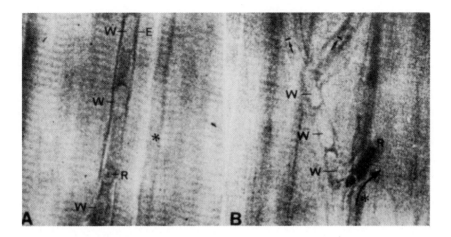

Fig. 3. Leukocyte plugging of skeletal muscle capillaries in cat during hemorrhagic shock. A. Asterisk: empty capillary, W: leukocytes, E: endothelial cell nucleus protruding into the lumen, R: erythrocytes. B. Leukocyte plugging at a capillary branching. Arrow with asterisk denotes direction of persistent flow [from Bagge et al, 1980].

number of leukocytes lodged in the alveolar capillaries in shock as compared to controls. Further, Wilson observed that the leukocytes seemed to be larger and more spherical in the bled animals. As discussed earlier such a morphological change of the cells might be accompanied by a reduced deformability, further augmenting the tendency of the leukocytes to become arrested in narrow vascular sections.

Besides capillary plugging by leukocytes another important feature of hemorrhagic shock is the adherence of large numbers of leukocytes to the vascular walls, primarily in the postcapillary venules [see eg, Amundson et al, 1980a]. The significance of leukocyte adherence, whether it occurs in arterioles or in venules, is that the stiff leukocytes protrude into the lumen and hence constitute a hindrance to the flow of the blood. According to Lipowsky et al [1980] wall-adherent leukocytes may cause a substantial increase—as much as 70% of the local viscous resistance after hemorrhage.

In a recent study Braide et al [1983] analysed the effect of leukocytes on the microcirculation in skeletal muscle under different pressure-flow conditions. These investigators used the isolated rat hindquarter preparation and isogravimetric technique to determine the changes in vascular resistance and capillary filtration coefficient (CFC) occurring after the infusion of a bolus of leukocytes. As could be predicted from the earlier studies in vivo under normal conditions the leukocytes infusions had very small or no effects on the vascular resistance or CFC when the hindquarters were perfused at flows normal for resting skeletal muscle. In contrast, at flows corresponding to those found in skeletal muscle in hemorrhagic shock, the infusion of leukocytes (in numbers per ml corresponding to those found in whole blood) caused a significant increase of the precapillary resistance, and a corresponding decrease of the CFC. Immediately upon infusion of the leukocytes the precapillary resistance increased as much as 70% but was then reduced to a constant level of about 20% above the control values. This sustained increase of the resistance was found to be positively correlated to the number of leukocytes remaining in the vascular bed. Attempts to reduce the vascular resistance by flushing the vascular bed by elevation of the perfusion pressure were only partially successful. This finding is in agreement with the result of Amundson et al [1980b], who found that restoration of blood volume and arterial pressure in bled animals could not, at least not within an observation period of 2 hours, normalize the distribution of the blood flow on the capillary level in skeletal muscle. At the present time the ultimate fate of the leukocytes lodged in the capillaries is entirely unknown. For reasons stated above it is not probable that they can return to the circulation by actively moving through the capillaries; a more likely development is that they undergo lysis after some time. The leukocytes adhering to the vascular walls of the venules may, on the other hand, disappear from the circulation by migrating through the endothelium into the tissues. This process takes several minutes, maybe half an hour, and in the acute phase treatment of hemorrhagic shock it may therefore be advisable—in order to reduce the flow-impairing effect of the wall adherent leukocytes—to administer drugs which can prevent adhesion of the leukocytes to the endothelium and also, ideally, break the binding forces between the leukocytes and the endothelium. Since a low pH promotes leukocyte adhesion [eg, Sung et al, 1982], a first step, which in fact is already taken in the treatment of shock but for other (metabolic) reasons, should be to abolish local tissue acidosis.

ACKNOWLEDGMENTS

Sponsored by grants from the Swedish Medical Research Council (12X-0063) and the Göteborg Medical Society.

REFERENCES

Adell R, Skalak R, Branemark P-I (1970): A preliminary study of rheology of granulocytes. Blut 21:91–105.

Amundson B, Jennische E, Haljamäe H (1980a): Correlative analysis of microcirculatory and cellular metabolic events in skeletal muscle during hemorrhagic shock. Acta Physiol Scand 108:147–158.

Amundson B, Jennische E, Haljamäe H (1980b): Skeletal muscle microcirculatory and metabolic effects of whole blood, Ringer's acetate and dextran 70 infusions in hemorrhagic shock. Circ Shock 7:111–120.

Applegren KL (1972): Capillary transport in relation to perfusion pressure and capillary flow in hyperemic dog skeletal muscle in shock. Eur. Surg. Res 8:311–320.

Asano M, Branemark P-I, Castenholz A (1973): A comparative study of continuous qualitative and quantitative analysis of microcirculation in man. Microchymography and microphoto-electric pletysmography applied to microvascular investivation. Adv Microcirc 5:1–31.

Baeckström P, Folkow B, Kovách AGB, Löving B, Öberg B (1971): Evidence of plugging of the microcirculation following acute hemorrhage. In Ditzel J, Lewis DH (eds): 6th Europ. Conf. Microcirculation. Aalborg 1970. Basel: Karger, pp 16–22.

Bagge U (1975): White blood cell rheology. Experimental studies on the rheological properties of white blood cells in man and rabbit and in an *in vitro* micro-flow system. Dissertation. Göteborg, Sweden: University of Göteborg.

Bagge U, Johansson BR, Olofsson J (1977a): Deformation of white blood cells in capillaries. A combined intravital and electron microscopic study in the mesentery of rabbits. Adv Microcirc 7:18–28.

Bagge U, Skalak R, Attefors R (1977b): Granulocyte rheology. Experimental studies in an *in vitro* micro-flow system. Adv Microcirc 7:29–48.

Bagge U, Brånemark P-I (1977): White blood cell rheology. An intravital study in man. Adv Microcirc 7:1–17.

Bagge U, Amundson B, Lauritzen C (1980): White blood cell deformability and plugging of skeletal muscle capillaries in hemorrhagic shock. Acta Physiol Scand 108:159–163.

Bessis M (1973): Living Blood Cells and their Ultrastructure. New York: Springer-Verlag, p 491.

Braide M, Amundson B, Chien S, Bagge U (1983): Quantitative studies on the influence of leukocytes on the vascular resistance in a skeletal muscle preparation. Microvasc Res (In press).

Brånemark P-I (1971): Intravascular Anatomy of Blood Cells in Man. Basel: S.Karger.

Brånemark P-I, Lindström J (1963): Shape of circulating blood corpuscles. Biorheology 1:139–142.

Dahlberg B (1979): Blood-tissue solute exchange in skeletal muscle during shock and trauma. Acta Physiol Scand [Suppl] 472, 107:1–82.

Eriksson E, Lisander B (1972): Low flow states in microvessels of skeletal muscle in cat. Acta Physiol Scand 86:202–210.

Eriksson E, Myrhage R (1972): Microvascular dimensions and blood flow in skeletal muscle. Acta Physiol Scand 86:211–222.

Folkow B, Neil E (1971): Circulation. Oxford: Oxford University press, p 37.

Fung YC, Tsang WCO, Patitucci P (1981): High-resolution data on the geometry of red blood cells. Biorheology 18:369–385.

Illig L (1957): Capillar "contractilität," capillary "sphincter" and zentralkanäle". Klin Wschr 35:7–22.

Kessler M, Höper J, Krumme BA (1976): Monitoring of tissue perfusion and cellular function. Anesthesiology 45:184–197.

Krogh A (1922): The Anatomy and Physiology of Capillaries. New York: Hafner Publ. Co., p 16. (Ed. of 1959).

Lichtman MA, Weed RI (1972): Alteration of the cell periphery during granulocyte maturation: Relationship to cell function. Blood 39:301–316.

Lichtman MA (1973): Rheology of leukocytes, leukocyte suspensions and blood in leukemia. J Clin Lab Invest 52:350–358.

Lipowsky HH, Usami S, Chien S (1980): *In vivo* measurements of "apparent viscosity" and microvessel hematocrit in the mesentery of the cat. Microvasc. Res 19:297–319.

Lund N, Ödman S, Lewis DH (1980): Skeletal muscle oxygen pressure fields in rats. A study of the normal state and the effects of local anesthetics, local trauma and hemorrhage. Acta Anaesthesiol Scand 24:155–160.

Mellander S, Lewis DH (1963): Effect of hemorrhagic shock on the reactivity of resistance and capacitance vessels and on capillary filtration transfer in cat skeletal muscle. Circ Res 13:105–118.

Miller ME, Myers KA (1975): Cellular deformability of the peripheral blood polymorphonuclear leukocyte: Method of study, normal variation and effects of physical and chemical alterations. J Reticuloendothel Soc 18:337–345.

Nicoll PA, Webb RL (1946): Blood circulation in the subcutaneous tissue of the living bat's wing. Ann NY Acad Sci 46:697–711.

Nobis U, Gaehtgens P (1980): Effect of white blood cells (WBC) on blood rheology in narrow tubes. Microvasc Res 19:395–396 (abstract).

Nobis U, Pries AR, Gaehtgens P (1982): Rheological mechanism contributing to WBC margination. In Bagge U, Born GVR, Gaehtgens P (eds): "White Blood Cells. Morphology and Rheology as related to Function." The Hague: Martinus Nijhoff Publishers, pp 57–65.

Palmer AA (1959): A study of blood flow in minute vessels of the pancreatic region of the rat with reference to intermittent corpuscular flow in individual capillaries. Q J Exp Physiol 44:149–159.

Palmer AA (1967): Platelet and leukocyte skimming. Bibl Anat 9:300–303.

Rand RP, Burton AC (1964): Mechanical properties of the red cell membrane. I. Membrane stiffness and intracellular pressure. Biophys J 4:115–135.

Sandison JC (1932): Contraction of blood vessels and observations on the circulation in the transparent chamber in the rabbit's ear. Anat Rec 54:105–127.

Schmid-Schönbein GW, Shih YY, Chien SS (1980a): Morphometry of human leukocytes. Blood 56:866–875.

Schmid-Schönbein GW, Usami S, Skalak R, Chien S (1980b): The interaction of leukocytes and erythrocytes in capillary and postcapillary vessels. Microvasc Res 19:45–70.

Schmid-Schönbein GW, Sung K-LP, Tözeren H, Skalak R, Chien S (1981): Passive mechanical properties of human leukocytes. Biophys J 36:253–256.

Sung K-LP, Schmid-Schönbein GW, Skalak R, Schuessler GB, Usami S, Chien S (1982): Influence of physiochemical factors on rheology of human neutrophils. Biophys J 39:101–106.

Vejlens G (1938): The distribution of leukocytes in the vascular system. Acta Pathol Microbiol Scand [Suppl]33:11–239.

Wilson JW (1972): Leukocyte sequestration and morphologic augmentation in the pulmonary network following hemorrhagic shock and related forms of stress. Adv Microcirc 4:197–232.

White Cell Mechanics: Basic Science and
Clinical Aspects, pages 295–306

The Relationship of Excessive White Cell Accumulation to Vascular Insufficiency in Patients With Leukemia

Marshall A. Lichtman

*Hematology Unit, Department of Medicine, University of Rochester
School of Medicine and Dentistry, Rochester, New York 14642*

In some cases of human leukemia, immature lymphocytes or granulocytes accumulate in the blood in extraordinary concentrations. In patients
with leukemia and hyperleukocytosis, clinical syndromes have been
observed that result from impaired circulation to the eye, the central
nervous system, the lungs, or the penis [1]. The occurrence of such organ
dysfunction referred to as the "hyperleukocytic syndrome" is correlated
with (1) the white cell count, that is, a greatly increased fractional volume
of blood composed of leukocytes; (2) the type of the leukocyte, that is,
granulocytic cells are more pathogenic of the hyperleukocytic syndrome
than are lymphocytic cells; and (3) the stage of maturation of the
leukocyte, that is, accumulation of immature (blast) cells is more prone to
result in symptoms than is an accumulation of more mature cells.

The whole blood viscosity measured in a viscometer is not increased in
most patients who have the hyperleukocytic syndrome and thus, the high
concentrations of immature leukocytes induce their deleterious effects by
altering the microcirculation in the organs involved [2].

WHITE CELL COUNT, LEUKOCRIT, AND THE RHEOLOGIC FEATURES OF LEUKOCYTES

Distribution of White Cell Counts in Leukemia

The distribution of blood white cell counts among patients with the four
major types of leukemia, acute myelogenous (AML), acute lymphocytic

(ALL), chronic myelogenous (CML), and chronic lymphocytic (CLL) leukemia at the time of their initial presentation is shown in Figure 1. Patients with AML and ALL have a similar distribution of white cell counts. About 10% of patients in these two groups have white cell counts above 100,000/μl and 5% have counts above 200,000/μl. Although patients with CLL have higher cell counts than subjects with AML or ALL, cell volume of CLL lymphocytes is smaller than either AML or ALL cells and the distribution of the true packed leukocyte volume (true leukocrit), calculated from the product of the total leukocyte count and mean cell volume measured in each of the three types of leukemia, is

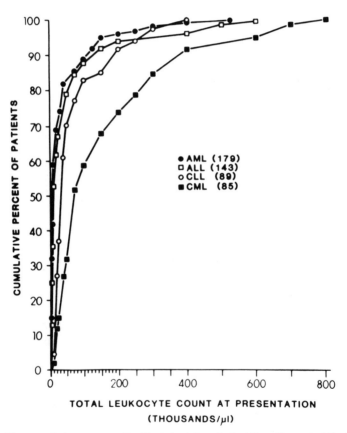

Fig. 1. The cumulative percent of subjects by interval of white cell count at the time of diagnosis is shown for subjects with acute myelogenous (AML), chronic myelogenous (CML), acute lymphocytic (ALL), and chronic lymphocytic (CLL) leukemias (reproduced from [1] with the permission of Grune and Stratton).

nearly identical (not shown). Patients with CML have higher leukocyte counts and higher leukocrits than do patient with the other major types of leukemia.

Leukocrit as a Function of White Cell Count

In order to produce a true leukocrit greater than 10 ml/dl, the concentration of leukocytes must increase to more than 200,000/μl, depending on the mean cell volume of the populations (Fig. 2). Leukemic lymphocytes (well-differentiated type) have a mean cell volume that ranges from 190 to 250 μm^3, leukemic lymphoblasts from 250 to 350 μm^3, and leukemic myeloblasts from 350 to 450 μm^3. Mean cell volumes are about 200 μm^3 in CLL, about 300 μm^3 in ALL, about 400 μm^3 in AML, and about 450 μm^3 in CML. The latter mean cell volume is increased somewhat as compared to AML because of the larger volumes of progranulocytes and early myelocytes, which are features of CML.

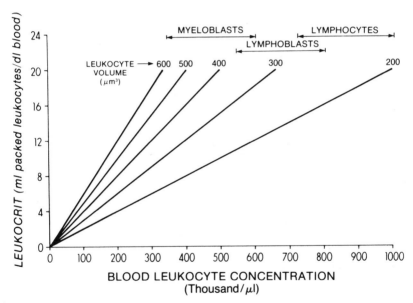

Fig. 2. The packed leukocyte volume (leukocrit) of blood on the ordinate is a function of the blood leukocyte count on the abscissa. The relationship of the true leukocrit to the leukocyte count is a function of the mean cell volume of the leukocyte population. Isovolumic diagonals are shown in the figure for different cell types. The true leukocrit on the oridinate is about 0.7 of the leukocrit observed using a microcytocrit centrifuge to sediment whole blood because of the poor deformability of leukocytes and resulting plasma trapping (reproduced from [1] with the permission of Grune and Stratton).

The Relationship of Erythrocrit to Leukocrit

In the chronic leukemias, both myelogenous and lymphocytic, there is an inverse relationship of erythrocrit with leukocrit. As leukocrit rises, erythrocrit falls. In CML and CLL, the erythrocrit falls somewhat more than the leukocrit rises (~1.3 ml/dl:~1.0 ml/dl), especially at high leukocrits. Thus, the total cytocrit is usually less than 45 ml/dl. In AML or ALL anemia is more severe at the white cell counts seen with hyperleukocytosis and thus the total cytocrit is less than that observed in hyperleukocytic chronic leukemias.

The Effect of Leukocrit on Viscosity

The viscosity of blood increases logarithmically as the fractional volume of red cells (erythrocrit) increases. Leukocytes are less deformable than are red cells and viscosity of blood increases more rapidly as the fractional volume of leukocytes (leukocrit) increases (Fig. 3). Although the viscosity of white cell suspensions is greater than that of red cell suspensions at equivalent fractional volumes, the differences are relatively small until the observed leukocrit (ie, that measured in a high-speed microcytocrit centrifuge) exceeds 15 ml/dl. The observed leukocrit is about 1.5 times the true leukocrit because the poorly deformable leukocytes are separated by a significant volume of trapped plasma. The true leukocrit can be estimated from the product of the cell packing factor (0.7) and the observed leukocrit. The true erythrocrit and observed erythrocrit are very similar.

CLINICAL FEATURES OF THE HYPERLEUKOCYTIC SYNDROME

The effects of enormous excesses of blood leukocytes occur in the circulation of the eye, the central nervous system, and especially in the brain, lung, and penis [3–29]. Signs of the syndrome are listed in Table I. Sudden death as a result of intravascular hemorrhage and irreversible neurologic damage has occurred. The ill effects of hyperleukocytosis on brain function have been reversible within hours after initiating a decrease in the leukocyte count by therapeutic leukapheresis.

Cytotoxic drugs administered to patients with hyperleukocytic leukemia may precipitate symptoms or signs that mimic the hyperleukocytic syndrome, perhaps as a result of the further loss of deformability of drug-injured leukemic blast cells or because of the release of procoagulants from the cells [30,31].

The frequency of the hyperleukocytic syndrome varies significantly according to the type of leukemia. Table II provides the frequency of cases by type of leukemia as determined in a retrospective review of patients

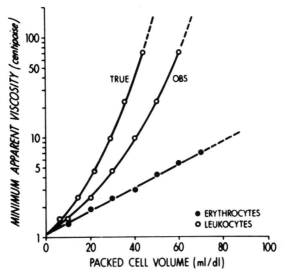

Fig. 3. The viscosity of suspensions of human leukocytes or erythrocytes in plasma. Cell suspensions were adjusted to specific packed cell volumes. The observed leukocrit (OBS) was not corrected for trapped plasma. True leukocrit (TRUE) was corrected for trapped plasma by multiplying the packing factor (0.7) and the observed leukocrit. Viscosity was measured in a cone-in-plate viscometer at different rates of shear. The minimum apparent viscosity represents the value derived from the square of the slope of the regression line of the square root of shear stress on the square root of shear rate, between rates of 2 and 200 sec^{-1} (reproduced from [1] with permission of Grune and Stratton).

records at the University of Rochester Medical Center. The occurrence of the syndrome correlates with the proportion of cases of AML or ALL with leukocyte counts above 200,000/μl and of cases of CML with counts above 350,000/μl. Patients with CLL do not appear to be predisposed to the syndrome even at very high white cell counts and this is probably a reflection of the difference in the intrinsic features of well-differentiated lymphocytes, which makes them less of an impediment in the microcirculation.

PATHOGENESIS OF THE HYPERLEUKOCYTIC SYNDROME

The precise events in the pathogenesis of the hyperleukocytic syndrome are not understood since the effects of leukemic cells on the flow characteristics of the microcirculation have not been thoroughly studied. Table III contains factors thought to be responsible for the clinical effects of large concentrations of leukemic leukocytes.

TABLE I. Clinical Features of the Hyperleukocytic Syndrome

Ocular
  Visual blurring, retinal vein distention,
  papilledema, retinal hemorrhages or exudates
Nervous
  Dizziness, tinnitus, ataxia, stupor, delerium,
  intracranial hemorrhage
Pulmonary
  Tachypnea, dyspnea, hypoxia
Other
  Priapism
  Peripheral vascular insufficiency

TABLE II. Frequency of the Hyperleukocytic Syndrome[a]

Type of leukemia	Total cases of syndrome
CML (85)	13 (15%)
AML (179)	9 (5%)
ALL (143)	6 (4%)
CLL (89)	1 (1%)

[a]Data derived from review of cases of leukemia at the University of Rochester Medical Center from 1970 to 1981. The number of patients with each type of leukemia is in parentheses.

TABLE III. Factors in the Pathogenesis of the Hyperleukocytic Syndrome

1. Increased viscosity in microcirculation
2. Stasis
3. Leukocyte aggregates
4. Leukocyte microthrombi
5. Oxygen steal and hypoxia
6. Microvascular invasion

An increase in blood viscosity as judged by viscometry is uncommon in hyperleukocytic leukemia except in a small proportion of cases of CML (Fig. 4). Since white cells influence blood viscosity in about the same manner as red cells until the observed leukocrit exceeds 15 ml/dl, and the fall in erythrocrit usually exceeds the rise in leukocrit, a rise above normal in the bulk viscosity of blood is uncommon in hyperleukocytic leukemia and is confined to cases of CML. In moderately anemic subjects (erythrocrit less than 30 ml/dl), an increase in observed leukocrit greater than 20

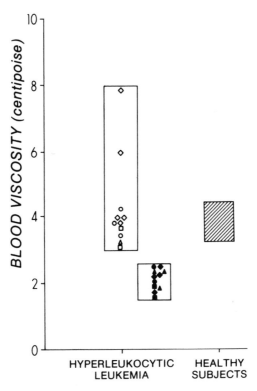

Fig. 4.   The viscosity of 11 consecutive subjects with hyperleukocytic leukemia and healthy subjects are compared. Diamonds represent five patients with CML with leukocyte counts of 210,000–620,000/μl. Circles represent three patients with AML with leukocyte counts of 160,000–230,000/μl. Squares represent two patients with ALL with leukocyte counts of 350,000 and 750,000/μl. The triangle represents a patient with CLL with a white cell count of 890,000/μl. The open symbols represent the patients blood. The closed symbols represent the blood of the patients with leukemia after the white cells have been removed but with the erythrocrit unchanged in autologous plasma (reproduced from [1] with the permission of Grune and Stratton).

ml/dl is required to increase the bulk viscosity of blood. Viscosity of blood in hyperleukocytic patients is greater than that observed in subjects with equivalent degrees of anemia and this could decrease the efficiency of oxygen transport at a given erythrocrit in hyperleukocytic patients as compared to patients with anemia in whom blood viscosity is less than normal [2].

Viscosity of blood in the microcirculation may play a role in the development of the hyperleukocytic syndrome. Viscosity in the microcirculation is a function of plasma viscosity and the deformability of individual cells in precapillary, capillary, or postcapillary vessels. Leukocytes, if they enter these vessels, should cause an increase in viscosity in small channels [32].

Flow in microchannels will fall if the diameter of poorly deformable white cells approaches that of the channel. The flow patterns of leukemic leukocytes in the microcirculation have not been studied. What vessels leukemic cells enter, and what their effects may be, can only be surmised. In vitro studies using micropore filters and micropipettes suggest that leukemic blast cells have properties that could lead to occlusion if the vessel caliber is less than 75% of the diameter of the leukemic cell [2,33]. Also, the frequency of leukocytes entering microchannels could be much greater than expected as a function of the markedly increased white cell counts. Figure 5 depicts the difference in deformability of leukemic

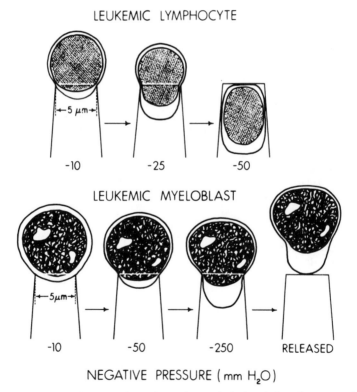

LEUKEMIC LYMPHOCYTE

-10   -25   -50

LEUKEMIC MYELOBLAST

-10   -50   -250   RELEASED

NEGATIVE PRESSURE (mm H₂O)

Fig. 5.   A schematic drawing of the shape change in leukemic lymphocytes and myeloblasts as they enter a 5-μm-diameter micropipette at applied negative pressures up to 250 mm H$_2$O (reproduced from [2] with the permission of the Rockefeller Press).

lymphocytes from patients with CLL and leukemic myeloblasts from patients with AML. Figure 6 depicts the difference in filterability of leukemic lymphocytes from patients with CLL and leukemic myeloblasts from patients with AML.

Leukemic cells consume oxygen at a rate that could contribute to a decreased concentration locally, if flow was compromised [33]. By competing with tissue cells in areas of obstructed flow, oxygen tension and thereby supply can be decreased (Fig. 7).

In some cases, the invasiveness of leukemic blast cells may cause a loss of integrity of the vascular wall. This effect has been thought to account for episodes of intracerebral hemorrhage in hyperleukocytic patients [6,7].

Histopathologic studies have shown that thrombi initiated by blast cell aggregates can lead to occlusion of small veins in the lung, brain, or other sites and that these findings are a function of white cell count and type of leukemia (myelogenous > lymphocytic) [10].

The hyperleukocytic syndrome, on the one hand, provides evidence for an uncommon but important clinical disorder that is directly related to the accumulation in the blood of immature, poorly deformable leukemic leukocytes. Its infrequency in the face of marked increases in normal or

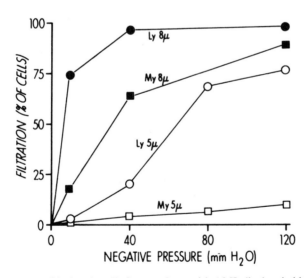

Fig. 6. The percent of leukemic cells from patients with AML (leukemic blast cells) and CLL (leukemic lymphocytes) that was filtered through 5- or 8-μm-pore-diameter filters as a function of the negative pressure applied to the filters is shown. The mean cell diameter for CLL lymphocytes is approximately 7.5 μm and of AML blasts approximately 8.5 μm.

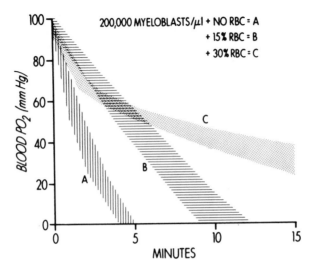

Fig. 7. The blood oxygen tension (PO₂) as a function of time in a closed system containing blood with 200,000 leukemic myeloblasts per microliter and either (A) no red cells or (B) a fractional volume of 15% red cells or (C) a fractional volume of 30% red cells. RBC, red blood cells (reproduced from [33] with the permission of Springer-Verlag).

near-normal cells (inflammatory leukemoid reactions), or of leukemic cells (most cases of leukemia with moderately elevated leukocyte counts), indicates that extreme deviations from normal are required to result in directly mediated leukocyte-induced vascular insufficiency.

REFERENCES

1. Lichtman MA, Rowe JM (1982): Hyperleukocytic leukemias: Rheological, clinical, and therapeutic considerations. Blood 60:279–283.
2. Lichtman MA (1973): Rheology of leukocytes, leukocyte suspensions and blood in leukemia: Possible relationship to clinical manifestations. J Clin Invest 52:350–358.
3. Thompson DS, Goldstone AH, Parry HF, Richard JDM (1978): Leukostasis in chronic myeloid leukemia. Br Med J 2:202.
4. Hild DH, Myers TJ (1980): Hyperviscosity in chronic granulocytic leukemia. Cancer 46:1418–1421.
5. Harris AL (1978): Leukostasis associated with blood transfusion in acute myeloid leukemia. Br Med J 2:11691.
6. Groch SN, Sayre GP, Heck FJ (1960): Cerebral hemorrhage in leukemia. Arch Neurol 2:439–451.
7. Freireich EJ, Thomas LB, Frei E, III, Fritz RD, Forkner CE, Jr (1960): A distinctive type of intracerebral hemorrhage associated with "blastic crisis" in patients with leukemia. Cancer 13:146–154.

8. Old JW, Smith WW, Grampa G (1955): Human lung in leukemia: Observations on alveolar capillary leukostasis with reference to pathologic physiology. Am J Pathol 3:605.

9. Resnick ME, Berkowitz RD, Rodman T (1961): Diffuse interstitial leukemic infiltration of the lungs producing the alveolar-capillary block syndrome. Am J Med 31:149–153.

10. McKee LC, Jr, Collins RD (1974): Intravascular leukocytic thrombi and aggregates as a cause of morbidity and mortality in leukemias. Medicine 53:463–478.

11. Ross JS, Ellman L (1974): Leukemic infiltration of the lungs in the chemotherapeutic era. Am J Clin Pathol 61:235–241.

12. Dearth JC, Fountain KS, Smithson WA, Burgert EO, Jr, Gilchrist GS (1978): Extreme leukemic leukocytosis (blast crisis) in childhood. Mayo Clin Proc 53:207–211.

13. Green RA, Nichols NJ, King EJ (1959): Alveolar-capillary block due to leukemic infiltration of the lung. Am Rev Resp Dis 80:895–901.

14. Geller SA (1971): Acute leukemia presenting as respiratory distress. Arch Pathol 91:573–576.

15. Vernant JP, Brun B, Mannoni P, Dreyfus B (1979): Respiratory distress of hyperleukocytic granulocytic leukemias. Cancer 44:264–268.

16. Karp DD, Beck JR, Cornell CJ, Jr (1981): Chronic granulocytic leukemia with respiratory distress. Arch Intern Med 141:1353–1354.

17. Frost T, Isbister JP, Ravich RBM (1981): Respiratory failure due to leukostasis in leukemia. Med J Aust 68:94–95.

18. Bloom R, Taviera DaSilva AM, Bracey A (1979): Reversible respiratory failure due to intravascular leukostasis in chronic myelogeneous leukemia. Am J Med 67:679–683.

19. Stirling ML, Parker AC, Keller AJ, Urbaniah SJ (1977): Leukapheresis for papilloedema in chronic granulocytic leukemia. Br Med J 2:676–677.

20. Preston FE, Sokol RJ, Lilleyman JS, Winfield DA, Blackburn EK (1978): Cellular hyperviscosity as a cause of neurologic symptoms in leukemia. Br Med J 1:476–478.

21. Eisenstaedt RS, Berkman EM (1978): Rapid cytoreduction in acute leukemia: Management of cerebral leukostasis by cell pheresis. Transfusion 18:113–115.

22. Suri R, Goldman JM, Catovsky D, Johnson SA, Wiltshaw E, Galton DAG (1980): Priapism complicating chronic granulocytic leukemia. Am J Hematol 9:295–299.

23. Ballas K, Kiesel JK (1979): Leukapheresis for hyperviscosity. Transfusion 19:787.

24. Heustis DW, Corrigan JJ, Jr, Johnson HV (1975): Leukapheresis of a five-year old girl with chronic granulocytic leukemia. Transfusion 15:489–490.

25. Lowenthal RM, Buskard NA, Goldman JM, Spiers ASD, Bergier N, Graubner M, Galton DAS (1975): Intensive leukapheresis as an initial therapy for chronic granulocytic leukemia. Blood 46:835–844.

26. Lane TA (1980): Continuous flow leukapheresis for rapid cytoreduction in leukemia. Transfusion 20:455–457.

27. Carpentieri U, Patten EV, Chamberlin PA, Young AD, Hitter ME (1979): Leukapheresis in a three-year old child with lymphoma in leukemic transformation. J Pediatr 94:919–921.

28. Kamen BA, Summers CP, Pearson HA (1980): Exchange transfusion as a treatment for hyperleukocytosis, anemia, and metabolic abnormalities in a patient with leukemia. J Pediatr 96:1045–1046.

29. Morse PH, McCready JL (1971): Peripheral retinal neovascularization in chronic myelocytic leukemia. Am J Opthalmol 72:975–978.

30. Lokich JJ, Moloney WC (1972): Fatal pulmonary leukostasis following treatment in

acute myelogenous leukemia. Arch Intern Med 130:759–762.

31. Hewlett RI, Wilson AF (1977): Adult respiratory distress syndrome (ARD) following aggressive management of extensive acute lymphoblastic leukemia. Cancer 39:2422–2424.

32. Palmer AA (1959): A study of blood flow in minute vessels of the pancreatic region of the rat with reference to intermittant corpuscular flow in individual capillaries. J Exp Physiol 44:149.

33. Lichtman MA, Kearney EA (1976): The filterability of normal and leukemic leukocytes. Blood Cells 2:491–506.

Index